Passport to Professional Numeracy

The following people have been involved in the production of the materials for this book:

Writing Team
Heather Cooke
Alan Graham
Jenny Houssart
John Mason

Editor
Eric Love

External Assessor
Pat Perks

Course Manager
Maggie Jenkins

Passport to Professional Numeracy

Arithmetic and Statistics for Teachers

HEATHER COOKE

David Fulton Publishers
London

in association with

Centre for Mathematics Education

David Fulton Publishers Ltd
Ormond House, 26–27 Boswell Street, London WC1N 3JZ

www.fultonpublishers.co.uk

First published in Great Britain by David Fulton Publishers 2001

Copyright © The Open University 2001

British Library Cataloguing in Publication Data
A catalogue record for this book is available from the British Library

ISBN 1–85346–789–8

All rights reserved. No part of this publication may be reproduced, stored in a retrieval system or transmitted, in any form, or by any means, electronic, mechanical, recording or otherwise, without the prior permission of the publishers.

The publishers would like to thank Kate Williams for copy-editing and Sophie Cox for proofreading this book.

Typeset by Elite Typesetting Techniques, Eastleigh, Hampshire
Printed in Great Britain by The Cromwell Press Ltd, Trowbridge, Wilts.

Contents

Acknowledgements	vii
Chapter 1 – Using this Book	1
Introduction	1
Working on tasks	2
Getting stuck	2
Comments	4
Identifying your needs	4
Active learning	5
Chapter 2 – Arithmetic	7
Using and representing different types of numbers	7
Names and numbers	7
Discrete and continuous	8
Measurements	11
Fractions, decimals and percentages	12
Fractions	12
Fractions and decimals	13
Combining fractions	14
Percentages	14
Calculating percentages	15
Converting to percentages	16
Combining percentages	18
Percentage increases/decreases	18
Ratios and proportions	20
Sorting out language	20
Manipulating comparisons	24
Calculating strategies	26
Calculating mentally	31
Calculating mentally with percentages	32

Mental calculations in context	33
Written methods	36
Calculators	37
Spreadsheets	42

Chapter 3 – Measures and Formulas — 45

The nature of measures and measurement	45
Using measurements in calculations	47
Units	48
The metric system	48
Converting units	49
Converting between imperial and metric units	51
Using formulas	56
Translating formulas	56
Using formulas in spreadsheets	59

Chapter 4 – Statistics — 65

Introduction	65
Grouped data	69
Some statistical terminology	69
Statistical summaries	72
Averages	73
Spread	75
Boxplots	77
Other statistical summaries	83
Representing and interpreting data in tables and graphs	83
Tables	83
Frequencies	91
Cumulative frequency	98
More on charts and graphs	101
Bar charts	103
Histograms	106
Pie charts	109
Line graphs	114
Scatter graphs (scatterplots)	116
Using performance indicators and benchmarking data	119

Resources — 123

Index — 125

Acknowledgements

DfEE Standards Unit for permission to use data from Autumn Package (1999 and 2000).

Thanks to all who have contributed by listening, trying examples, and commenting.

CHAPTER 1

Using this Book

Introduction

Schools are now required to handle a great deal of numerical information – some for strategic management and evaluation, but mostly for day-to-day planning and monitoring. This book is designed to help you to use and to make more sense of this information. In particular, teachers in training are required to demonstrate their skills and knowledge of basic arithmetic and statistics in school contexts to obtain qualified teacher status (QTS). This book provides the means to revise or acquire those skills.

Many people, even those who have successfully passed mathematics examinations, lack confidence in their own knowledge and understanding. *Passport to Professional Numeracy* is written with such people in mind. Much of the numerical information dealt with in schools is statistical, and that is where the emphasis of this book lies. However, to understand such information it is necessary to be confident with arithmetic, measures and formulas, and the first two sections of the book are concerned with these aspects. The sections are designed to stand alone so you can organise your study to meet your own particular needs.

> **Task 1   Getting an overview**
>
> Quickly flick through the book to get a feel for how it is organised. Look at the Contents page, and index at the back. Perhaps quickly read through a few 'tasks' and some text to get a flavour of the way the book is written. What do you notice?
>
> Start to think about how *you* might use this book.

Comment
The book is designed for self-study and so uses activities as a teaching medium. You may be unfamiliar with this active learning approach. Every learner is different and so there is no best way to learn but the methods employed here have been found to be effective for people working on their own.

After you have examined the contents and thought about your own existing knowledge, possibly with the help and advice of others, you may eventually decide either to work through all the sections in the book or to concentrate on particular sections or subsections. The choice is yours.

Working on tasks

The topics in this book have been chosen to cover the Teacher Training Agency (TTA) Numeracy Skills QTS requirements, but they also go beyond those. The approaches have been devised to assist you in understanding the purposes of the mathematical ideas and not simply to explain how to carry them out. Working on the tasks is an essential part of the learning process; to gain the most from this book you will need to carry them out.

The tasks:
- ask questions to get you thinking about a particular topic;
- ask you to make notes on how you solve particular problems;
- challenge your mathematical thinking;
- review what and how you are learning;
- practise skills and techniques.

You may need to think about how you tackle each task and for how long. For many of the activities there is no right or wrong answer.

Getting stuck

When you start to work on a task you may feel that you are not sure what to do or how to do it: you are 'stuck'. There are a number of strategies that can make this a positive experience rather than a negative one.

1. Acknowledge that you are stuck, relax and recognise that this is a learning opportunity! Different people develop different strategies for dealing with being stuck. Whatever you do, don't panic.

2. Next try to identify exactly why you are stuck. This process is in effect identifying what you do *know* and what you *want*. Doing this can sometimes be enough for you to see a way of building a bridge between *know* and *want* ... and so become 'unstuck'.

3. Now try and do something about being stuck.

    - If the question seems too complicated or too general, try simplifying it in some way. For example, break it down into a sub-set of smaller problems, or rewrite it using simpler numbers or easier words.

    - If there does not seem to be enough information, list what else you think you need. (Some tasks may deliberately not have enough information.) Sometimes you may find that you do have the information but it was not in quite the form you expected.

    - Tell someone. In trying to explain you may find that you stress and ignore different parts of the problem and so view it in a new light. Even if there is no one around to help, just saying something out loud to yourself can help considerably; saying it 'in your head' is not as powerful.

    - Use the solution. Start to read the solution. (In this book they are in the Comments that follow each task.) You may only need to read a little before you can see what is needed and can continue on your own. Or you may need to work through the entire solution before light dawns.

    - If you are still stuck, still do not panic – you may need to take a break and do something quite different. Simply freeing your attention can unblock the problem.

    - If nothing seems to work, skip over the problem area for the moment and look at it again later.

When you have been 'stuck' and then manage to become 'unstuck', think about what happened. The way to make being stuck a positive experience is to notice not only what helped to get you going again, but also what led you to get stuck in the first place. This 'learning from experience' is then available to you for use in future situations.

Comments

At the end of each task is a 'Comment' section, which is part of the teaching text; where applicable, it gives numerical solutions and explanations. You may be tempted to look at the comments before doing any work on the task yourself but for effective learning you are advised to 'have a go' first. The Comment sections need to be worked on as much as the tasks themselves: active learning is just that – active! For example, when there is a numeric answer there may be different ways of arriving at it, and you may need to think about the best way for you.

Identifying your needs

Task 2 What do you need?

To help you to target your study time, try to identify what outcomes you need from studying this pack. Which of the following aspects do you want to improve your understanding and skills in?

- Arithmetic
    - Using different types of numbers
    - Fractions, decimals and percentages
    - Ratio and proportion
    - Calculating strategies
    - Using calculators and spreadsheets for calculations
- Measures and formulas
    - Different units of measurement
    - Converting between different units
    - Using weighted scores
- Statistics
    - Kinds of data
    - Summarising data by averages and spread
    - Representing and interpreting data in tables, graphs and charts
    - Using performance indicators and benchmarking

Comment

If you are studying the book in preparation for the QTS Numeracy Skills Test, it is worth trying the practice test questions that are available from the TTA website (see Resources section for address). This will enable you to target those aspects that are problematic for you, and enable you to make the most effective use of the study time you have available.

To help you in studying this book it is useful to organise your time. You will need time to work through the text and tasks, but you may need extra time to practise new skills. Also, new ideas take time to be absorbed, so you will need to allow for that. You might find it useful to organise your time, perhaps by making a list of what you intend to study and how long you intend taking; it may help to make your own study timetable.

You may find aspects of the book that go beyond your present needs, but it will be available to provide knowledge and understanding as you need it during your career.

Active learning

For whatever reason you are using this book, do try to be as active a learner as possible. The importance of the tasks has already been stressed, but you can also help your learning by making notes. The very act of writing notes or telling someone else, however briefly, can promote understanding. How you make notes is up to you – use the margins, stick in extra pages, use highlighters – but don't just read, be active!

CHAPTER 2
Arithmetic

Information is rarely used unless for a purpose and this is often to provide evidence in support of an argument. Where information is used as an aid to decision-making it usually needs to be processed or adapted in some way, perhaps by means of a further calculation or maybe with the addition of a graph. The scope for the misuse of information in these formats (graphs and calculations) is legendary. Indeed, maybe the best reason for studying mathematics is self-protection from those seeking to beat, cheat or mistreat you in some way.

A useful starting point is to have a clear overview of the basic ways in which information can be organised. There is more than one way of categorising information but the schemes introduced below are both well recognised and useful in helping you to grasp some of the do's and don'ts of information processing. These categories are developed more fully in Statistics, pp. 69–72.

Using and representing different types of numbers

Names and numbers

Perhaps the key idea in categorising information is the distinction between *names* and *numbers*. For example, when information is provided about a school you might see the name of the school, the name of the head teacher, the address, and so on; little in the way of numerical information here, but largely words or names. Numbers, on the other hand, crop up when you need to look at items such as the number of pupils on the roll, the annual expenditure, examination performance measures and so on. Sometimes numbers are used as names; for example telephone numbers. You can tell they are being used as names because there is no numerical significance to the order of telephone numbers in a list.

Discrete and continuous

A second important distinction that can be made is between quantities that can be counted and those that are measured. Quantities that you count, for example number of children in a family or the number of mathematics pupils gaining each GCSE grade, are called **discrete**. Discrete data are such that they can fall into separate categories (you cannot have a fraction number of children in a family). Contrast this with the measurement of a child's height: a child might be *any* height within a given range. Quantities that are measured are called **continuous**. As you will see later in Statistics (p. 70), this distinction between data that are discrete and those that are continuous is a key one for deciding how the data can be depicted graphically and what sorts of calculations can be applied.

Table 1 Examples of discrete data

Example	Why these data are discrete
1. Whether or not a pupil qualifies for free school meals	There are only two possible responses: yes and no
2. Which subjects are offered by a secondary school	The list (mathematics, PE, English, and so on) may be extensive but there are only a finite number of possibilities
3. Means by which pupils travel to school	The possibilities (on foot, by car, bus, bicycle and so on) cover only about six or seven options
4. School year groups	Years 1, 2, 3, 4 and so on form only a few categories
5. Children's shoe sizes	Sizes 1, $1\frac{1}{2}$, 2, $2\frac{1}{2}$ and so on form only a few categories

As you can see from Table 1, discrete data can take the form either of names (examples 1–3) or numbers (examples 4 and 5). For data to be continuous, they must satisfy two conditions: they must be numbers and they must be measured on a scale that can take an infinite number of possible values. Examples include a person's height and their age.

Note that though age, like most measures of time, is continuous, on a questionnaire most people will state their age rounded down to the full year below (someone who is 25 years and 7 months old will say that they are 25 years old), so age data may actually be recorded as discrete numbers.

> **Task 3 Discrete or continuous?**
>
> For each example in Table 2, make two ticks, the first indicating whether the values it can take are names or numbers and the second whether the measure is discrete or continuous.

Table 2 Discrete and continuous data

Example	Names	Numbers	Discrete	Continuous
1. Area of the classroom floor (m^2)				
2. The number of floor tiles required to cover a classroom floor				
3. The list of staff names in a school				
4. A child's reading age				
5. A list of sporting items ordered by PE staff				
6. The number of children on the school roll				
7. The makes of cars parked in the school car park				
8. Air temperature measured each day in a classroom				
9. Days of the week				
10. Class attendance on each day of the week				

Comment
You should have Table 3.

Table 3 Discrete and continuous data

Example	Names	Numbers	Discrete	Continuous
1. Area of the classroom floor (m²)		✓		✓
2. The number of floor tiles required to cover a classroom floor		✓	✓	
3. The list of staff names in a school	✓		✓	
4. A child's reading age		✓		✓
5. A list of sporting items ordered by PE staff	✓		✓	
6. The number of children on the school roll		✓	✓	
7. The makes of cars parked in the school car park	✓		✓	
8. Air temperature measured each day in a classroom		✓		✓
9. Days of the week	✓		✓	
10. Class attendance on each day of the week		✓	✓	

As you may already have spotted, only numerical data can have the possibility of being continuous; names are *always* discrete. This point is reinforced in Task 4.

Task 4 Names, numbers, discrete and continuous

The table below brings together your grasp of the names/numbers distinction and the discrete/continuous distinction.

The word *always* has been written in the first cell of the table because data that are names are always discrete. Complete the other three blank cells using an appropriate word in each from the list: always; sometimes; never.

	Names	Numbers
Discrete	always	
Continuous		

Comment

You should have come up with the following table:

	Names	Numbers
Discrete	always	sometimes
Continuous	never	sometimes

Measurements

The central purpose of measurement is comparison. The measurements that a school needs range from the size of chairs needed to match the size of pupils through to complex performance indicators that enable year-on-year comparison internally and with other schools.

Many of the numbers met in schools are measurements of one kind or another. The school day is determined by a timetable; the number of pupils in a particular room may be decided by its size; staff numbers and resources are limited by the budget; enrolment can be affected by published performance figures. How measurements can be used depends on the particular purpose; but factors such as type of measure, degree of accuracy and the units used also need to be considered. Issues involving physical measurements are considered in more detail in Measures and Formulas (pp. 45–63).

Fractions, decimals and percentages

Fractions, decimals and percentages can all be used to express parts of a whole. It is possible to use any of these forms and to convert from one to the others. So, for example, $\frac{1}{2}$ can also be expressed as 0.5 or as 50%. Fractions and percentages in particular are often used in describing pupil and school performance.

- Of 12 pupils who sat tests from a school of 82 pupils, $\frac{3}{4}$ achieved Level 5.
- In a school in which almost 70% of pupils were entitled to free meals, more than 80% reached the expected level in English.
- Three-quarters of all schools improved their overall score, and a third improved for the second year running.

> **Task 5 Why use fractions and percentages?**
>
> Read through the list of examples above again. Why are fractions and percentages used?

Comment
The fractions and percentages are being used to make comparisons. Note that percentages are also fractions but use a different notation (so 70%= $\frac{70}{100}$).

Fractions

There are three main ways in which fractions are used.
- Parts of a whole: the fraction $\frac{4}{5}$ can mean four-fifths of a 'whole'. Any fraction can be thought of as signalling an operation to be performed on some (often unspecified) unit. When fractions are being used as comparisons (as four-fifths of the pupils) then there is a stated or implied quantity, in this case, 'pupils' meaning 'total number of pupils'.
- A division: the fraction $\frac{4}{5}$ can also mean 4 divided by 5, and that is how to convert the fraction to a decimal (said, divide 4 by 5 to get 0.8).
- A ratio: the fraction $\frac{4}{5}$ can also signal a ratio, that is, a comparison: four of these to five of those.

> **Task 6   Fractions of a number**
>
> 1. In a class of 27 children, $\frac{1}{3}$ will be playing football, $\frac{1}{3}$ bench ball and $\frac{1}{3}$ netball. How many children are in each group?
>
> 2. In a class of 28 children, $\frac{3}{4}$ of the class want to go on a residential visit. How many children is this?

Comment
Both calculations involve division and the second also involves multiplication. To explain a possible method it is worth considering the different parts of a fraction. The **denominator** of a fraction (the bottom number) shows what size part is being expressed, while the **numerator** (the top number) indicates how many of these parts there are.

1. Finding $\frac{1}{3}$ of 27 is straightforward because the numerator of the fraction is 1, so we just need to divide by the denominator, giving $27 \div 3 = 9$ or put another way $\frac{27}{3} = 9$.

2. To find $\frac{3}{4}$ of 28, first divide 28 by 4 to find one-quarter of 28, giving an answer of 7. Then find three-quarters by multiplying by 3 to give an answer of 21. In other words, divide by the denominator then multiply by the numerator.

Fractions and decimals

Because a fraction can be seen as a division, fractions can be converted to decimals by carrying out this division. So, for example $\frac{2}{5}$ (2 divided by 5) becomes 0.4 and $\frac{1}{4}$ becomes 0.25. Sometimes, as with $\frac{2}{3}$, the result is a recurring decimal, in the case of $\frac{2}{3}$, 0.666…. This is often expressed to a certain number of decimal places. For example $\frac{2}{3}$ to 2 decimal places (d.p.) would be written as 0.67 (because the second 6 has been rounded up). Decimals often are used because they suggest a greater accuracy than whole numbers, but that cannot always be assumed.

Converting decimals to fractions depends on understanding what the digits in a decimal represent. Because the first number after the decimal point represents tenths, 0.8 is equivalent to $\frac{8}{10}$ (or $\frac{4}{5}$). Similarly because the second number after the decimal point represents hundredths, 0.15 is equivalent to $\frac{15}{100}$ (or $\frac{3}{20}$).

Combining fractions

Suppose two-thirds of one school qualify for free school meals, while only a half of another school qualify. What fraction overall qualify? You certainly do not simply add the fractions. Rather, you remember to ask what they are fractions *of*, and compute the actual numbers.

> **Task 7 What fraction?**
>
> In one school of 300 pupils, two-thirds qualify for free school meals; in a second school with 78 pupils, a half qualify.
>
> 1. How many pupils qualify altogether? What fraction is this of the total number of pupils?
>
> 2. Write each of these fractions as a decimal.

Comment

1. The total number of pupils that qualify for free school meals is $\frac{2}{3}$ of 300 + $\frac{1}{2}$ of 78, that is 239. Thus the fraction is 239 out of 378 pupils, or $\frac{239}{378}$.

2. $\frac{2}{3}$ is 0.667; $\frac{1}{2}$ is 0.5; and $\frac{239}{378}$ is 239 divided by 378, which is 0.632. As you would expect this lies between the fractions for each school, but closer to that of the bigger school. It is much easier to compare decimals than fractions; although fractions are a convenient short-hand when using small numbers, decimals – or percentages – are usually easier to interpret.

Percentages

A percentage can be thought of as a fraction where the denominator is 100. So, for example, 30% is $\frac{30}{100}$ or $\frac{3}{10}$.

With percentages, 100% is 'the whole'. So if 87% of a class pass a test, 100% − 87% = 13% fail it. But when using percentages as a way of expressing a fraction of a whole, you need to be clear what the 'whole', the 100%, is. This is often a source of confusion and can frequently be misleading. For example, if claims are made about the percentage of unauthorised absences in a school, you need to be clear whether this is a percentage of all the absences, or a percentage of all pupil-days (the number of pupils multiplied by the number of days in the school year – normally 190; the number of registrations being 380). These different 'wholes' give very different percentage values.

Percentages are also commonly used to show increases or decreases, for example in prices, school attendance or examination passes. Such percentage increases or decreases are calculated as a percentage of the original. Knowing what the 'original' is can be as problematic as knowing the 'whole'. For example, it means very little to say that A level results for a school have gone up by 5%. This could simply be the total number of passes in the school one year, compared to the total number of passes the previous year. However, it could be the number of passes per student – a fairer comparison if student numbers are changing. Even then, there is an issue about which students are being included: all students who passed at least one A level; all those who sat the exam; all those eligible to sit the exam? It might even be that comparisons are being made on the basis of point scores based on grades, rather than simply on number of passes.

Calculating percentages

Because the denominator of a percentage is 100, it is most useful, especially when using a calculator, to think of percentages as decimals in disguise: 35% = 0.35. This means that finding a percentage of something is the same as multiplying by a decimal. So, for example, 27% of 243 is $0.27 \times 243 = 65.61$. When percentages correspond to simple fractions (e.g. 50%, 10%, 75%) it is often quicker to use simple mental methods. The following task has an example of each.

> **Task 8 Calculating percentages**
>
> Calculate the following:
>
> 1. A school gives the percentage of pupils entitled to free school meals as 17%. There are 396 pupils in the school. How many are entitled to free school meals?
>
> 2. There are 472 pupils in a school, and 75% of them achieve full attendance over a term. How many pupils is this?

Comment

There are several different ways of carrying out each of these calculations, but all are likely to make use of conversion from percentages.

1. Writing 17% as a decimal we have 0.17. Multiplying 0.17 by 396 gives 67.32. As the answer clearly has to be a whole number, this means 67 children are entitled to free school meals. This is a reminder that the 17% will have been rounded.

2. Because 75% is a simple fraction, one method is to convert 75% to $\frac{3}{4}$. Then $\frac{3}{4}$ of 472 can be found by dividing by 4 then multiplying by 3, to give 354.

Another method:

50% of 472 is 236, so 25% of 472 is 118.

Add these to give 75% of 472 is 354.

Knowing the fractional equivalents of common percentages such as 50%, 20%, and 25% can be useful as a quick mental method. Mental methods of carrying out simple percentage calculations are discussed further on p. 32.

Converting to percentages

It is not easy to compare fractions: which is the greater, $\frac{3}{5}$ or $\frac{7}{11}$? Comparing two percentages or two decimals is much more straightforward, so if you need to know the relative sizes of two fractions it makes sense to change them into decimals or percentages.

Because decimals and percentages are two ways of writing the same thing, fractions can be changed to decimals first and then to percentages.

For example, $\frac{3}{5}$ is changed to a decimal by finding $3 \div 5 = 0.6$. To turn this into a percentage you need to think of $\frac{3}{5}$ths *of* 100% (the 'whole'), that is $(\frac{3}{5} \times 100)\%$. So $(3 \div 5 \times 100)\% = 60\%$.

Similarly, $\frac{7}{11}$ is changed to a percentage by calculating $7 \div 11 \times 100$ and writing the solution with a percentage sign (63.6% to 1 d.p.).

> This amounts to a rule for converting a proportion (a fraction) to a percentage: multiply the fraction by 100 and write as a percentage.
>
> To convert a percentage to a fraction, write the percentage as a fraction (e.g. 60% = $\frac{60}{100}$ and look for a simpler equivalent fraction (e.g. $\frac{60}{100} = \frac{6}{10} = \frac{3}{5}$).

Task 9   Converting to percentages

Use a calculator to convert the following fractions to percentages. Where necessary, give answers to 1 decimal place.
1. $\frac{3}{8}$ 2. $\frac{4}{9}$ 3. $\frac{4}{5}$

Comment
1. 37.5%

2. 44.4%

3. 80%. Here $4 \div 5 = 0.8$, so changing to a percentage is $0.8 \times 100\% = 80\%$.

Task 10   Transfer

If 5500 out of a total of 9500 people employed by Training and Enterprise Councils (TECs) are transferred to the new councils, what (simplified) fraction and what percentage is this?

Comment
As a fraction, 5500 out of 9500 is $\frac{5500}{9500} = \frac{55}{95} = \frac{11}{19}$

To turn this into a percentage, $\frac{11}{19} = 11 \div 19 = (11 \div 19 \times 100)\% = 58\%$ to the nearest whole number.

Combining percentages

> ### Task 11 School meals
> In one school, 20% of the pupils are eligible for school meals, while in a nearby school 25% are eligible for school meals. What percentage are eligible in the two schools together?

Comment
You cannot answer this question unless you know the pupil numbers for each school. As with fractions, in order to combine percentages, you must make sure that they are proportions of the same thing. Here we are concerned with numbers of pupils: to compute the overall percentage we have to find the total number of pupils.

For example, suppose school A has 100 pupils and school B has 400. Then we have 20% of 100 + 25% of 400 eligible pupils, which is (20 + 100 =)120 out of 500, or 24%. Suppose the pupil numbers were the other way round. Then 20% of 400 + 25% of 100 is (80 + 25 =)105 out of 500, or 21%. It makes a difference!

Percentages are useful when there are large numbers involved, or when comparing proportions. They can be misleading when applied to small numbers. Consider the following:

> Of the 26 pupils who took the tests, only 15% reached Level 4 in English, 4% reached Level 4 in mathematics, and 4% reached Level 4 in science.

The number 26 is far too small to make a percentage figure useful, since a slight change in the percentage means an extra pupil (through rounding). There must have been considerable rounding, since 15% of 26 is 3.9 (4 pupils?), and 4% is 1.04.

Percentage increases/decreases

> ### Task 12 Price cut
> A book previously sold at £3.40 is now on offer for £2.85 from one supplier and with a 20% discount from another. Which is the better buy?

Comment
There are several ways this can be worked out. You could:

- calculate the percentage reduction from the first supplier;
- work out the actual price from the second supplier;
- calculate the cash reduction from both suppliers.

The first supplier is offering a £3.40 − £2.85 = £0.55 reduction. As a percentage this is ($\frac{0.55}{3.40}$ × 100)% = 16.2% to 1 d.p. This is a smaller percentage discount than that offered by the second supplier.

The second supplier discount is $\frac{20}{100}$ × £3.40 = £0.68, so the price is £3.40 − £0.68 = £2.72. (See pp. 32–33 for calculating percentages mentally.)

Whichever way you worked it out, the second supplier is offering the better deal.

In reality, other factors might affect your choice, for example postage and packing charges and speed of delivery.

Task 13 Improvement

One year 20 out of 32 pupils achieve at least Level 4 in Key Stage 2 Standard Assessment Tasks (SATs). The following year 20 out of 28 pupils achieve at least Level 4. By how much has the school improved its performance?

Comment
The first year, $\frac{20}{32}$ = 62.5% of the pupils achieved at least Level 4.

The second year $\frac{20}{28}$ = 71.4% achieved at least Level 4.

It is tempting to say that there has been a 8.9% increase (71.4% − 62.5% = 8.9%), but the two percentages are percentages of different sized groups of pupils and so cannot be combined in that way. Notice that the pupil numbers are again small here, and so percentage increases or decreases need to be viewed with caution.

However, it is now common practice in school performance data to refer to *percentage point* increases or decreases, so in this case the second year performance is an 8.9 percentage point improvement. The justification for this practice is that, in most schools, year groups are of similar size. Percentage point measures are not suitable for schools with large fluctuations in year group numbers, for example very small schools or ones with transient populations (traveller or services families).

> **Task 14   Explaining the difference**
> In your own words write down the difference between percentage increase and percentage point increase. Then explain it to someone else.

> **Strategies for avoiding common errors with fractions, decimals and percentages**
>
> - Be attentive to what 'whole' a fraction or a percentage refers to.
>
> - Understand that fractions and percentages can be inappropriate or misleading when applied to data involving small numbers.
>
> - Where fractions or percentages are applied the resulting decimal needs to be interpreted in context.
>
> - Be aware of the difference between percentage increase/decrease and percentage point increase/decrease (and the limitations of this measure).

Ratios and proportions

Sorting out language

The words *ratio, proportion, fraction* and *percentage* are all used in ordinary speech, but not always with their mathematical meaning.

Task 15   Recalling usage

Try to recall at least three contexts in which each of the words *ratio, proportion, fraction* and *percentage* is commonly used. Jot down some examples.

Comment

Note that we normally speak of a (specific) fraction *of* something, a percentage *of* something, a proportion *of* something, but a ratio *of* something *to* something:

'The ratio of girls to boys in a class is 18 to 16 (or equivalently 9 to 8).'
'Two out of three girls prefer to talk at break time rather than play a game.'
'The proportion of girls in the class is 18 out of 34 (or equivalently 9 out of 17 or nearly 1 in 2).'
'The girls who misbehave in school are a (small) fraction of the whole.'
'Three in four (three out of every four) pupils are late for school at least once in a term.'

Ratio is a means of comparing two (or more) quantities, as in girls to boys, or those who take school meals to those who do not, or those achieving Levels 4 or 5 on a Key Stage 2 SAT test to the total taking the test.

Whereas fractions, percentages and proportions are each of 'a whole', ratio is used when comparing two quantities, neither of which need be a whole. The significant feature of a ratio is that it measures a relationship.

Example

The staff–student ratio in a school means the ratio of number of staff to the number of students. In a particular school the staff–student ratio might be 1 to 25. In other words, you could assign each member of staff to 25 students and that would assign everybody.

Notice that there is no indication of the actual numbers of staff or of students. The school may have 200 students or 1000 students, or any number. Again, the ratio does not tell you anything about how the school organises its classes. It does not mean that each class has 26 students, and it certainly does not mean that if you talk to 26 people in the school that one will be a member of staff and the others students!

A common misuse of the notion of 'staff–student ratio' is to give the numbers the wrong way round. Because people like to have a large number first, and because to say 'student–staff ratio' is awkward, there is often confusion and people say 'the staff–student ratio is 25 to 1' when they mean that it is 1 to 25. The numbers in a ratio should always be given in the same order as the statement.

The shorthand way of writing ratios is to use a colon to replace 'to'. Thus, the staff–student ratio is $1:25$. Like fractions, it is usual to scale down ratios to the lowest whole numbers, so a ratio of $5:15$ would normally be given as $1:3$. If in doubt, think of picking berries or shelling peas: one for me and three for the pot, one for me and three for the pot, over and over, means a ratio of 1 to 3, written $1:3$.

Ratios are almost always written using integers. We do *not* write $3.5:4.5$, but rather $35:45$ or $7:9$.

Proportions are like fractions in that they assume a whole. Thus the proportion of children gaining Level 4 in their SATs will be the fraction of all those taking the SATs. The 'whole' here may be a particular class or all the children in a school who took the tests or all the children of that age group in the country: the context will usually make it clear.

Sometimes proportions are used in rather more indirect ways. For example, in 'The proportion of girls gaining Level 4 was higher than the proportion of boys' there are two different wholes. The number of girls gaining Level 4 is being compared with the total number of girls and the number of boys gaining Level 4 is being compared with the total number of boys. The reason this is done is because the numbers of girls and boys may not be equal, and so simply to say 'more girls than boys reached Level 4' would not tell us whether girls were performing better than boys (there might be more girls in the group).

Task 16 Ratio and proportion

In the situations below you are asked to work out a ratio (and simplify it if possible) or a proportion.

1. Of 35 chairs in a room, five are broken. What is the ratio of broken to unbroken chairs? What is the proportion of broken chairs, and proportion of unbroken chairs?

2. There are 64 half-day absences among 256 children. What is the ratio of half-day absences to the number of children? The term has 55 days. What is the proportion of child-day-absences to child-days? Which number is more useful for gauging absences?

Comment

1. The ratio of broken to unbroken chairs is $5:30$, or $1:6$. The proportion of unbroken chairs is 30 out of 35, or $\frac{30}{35}$, so $\frac{6}{7}$ of the chairs are unbroken. Therefore $\frac{1}{7}$ are broken. In quoting figures like these, there is usually a choice between accentuating the positive (proportion of unbroken chairs) or the negative (proportion of broken chairs).

2. The ratio of half-day absences to the number of children is $64:256$ or $1:4$. That is, there was one half-day absence for every four children. This ratio looks very worrying. However, it makes more sense to compare the number of absences to the possible number of child attendances, rather than to the number of children. For a term of 55 days, or 110 half-days, the possible number of attendances for 256 children is 110×256. So the ratio of absences to possible attendances is $64:28\,160$ which is $1:440$; a much less worrying ratio.

Task 17 Lunchtime

In a school of 240 pupils, if 40 go home for lunch, what proportion go home for lunch and what proportion stay at school?

Comment

40 out of 240, or $\frac{40}{240} = \frac{4}{24} = \frac{1}{6}$ go home.

200 out of 240, or $\frac{20}{24} = \frac{5}{6}$ stay.

Note that the proportion that 'do not' is '1 minus the proportion who do': $\frac{1}{6}$ do go home and $1 - \frac{1}{6} = \frac{5}{6}$ do not.

In Task 17 you were given the numbers and asked for proportions. Often you are given a proportion and some number and asked to work out other numbers. Suppose you are told the proportion who go home (say $\frac{1}{6}$), and how many of them there are (say 30). Can you find the number who stay at school? Or suppose you are told the proportion who go home (say $\frac{2}{5}$) and the school population (say 250). Can you find the number who stay for meals? These are common switches: previously you knew the numbers and wanted the proportion, now you know the proportion and some number and you want the rest of the numbers.

Whenever you encounter what seems like a switch on what you are used to, write down what you know, and lay it alongside a more familiar computation, if need be.

So $\frac{1}{6}$ stay at home, and these amount to 30. What number is 30 one-sixth of? Or put another way, $\frac{30}{what?} = \frac{1}{6}$. Either by 'thinking it through', or by scaling $\frac{1}{6}$ by 30 to get a 30 on the top, or by 'cross multiplying', we get $30 \times 6 = 180$ children all told. We want to find the number who stay for meals, which is either $180 - 30$ (subtract those who go home) or $\frac{5}{6}$ of 180; which is 150 by either method.

Similarly, if $\frac{2}{5}$ go home, then work out the number of children ($\frac{2}{5}$ of 250 = 100) and subtract from 250 to give 150. Alternatively, note that $\frac{3}{5}$ stay, so $\frac{3}{5}$ of 250 = 150, as before.

Manipulating comparisons

There is a temptation to combine ratios and proportions, sometimes appropriately, but sometimes not. In every case it is wise to ask oneself 'are these fractions or percentages of the *same thing*?'

Task 18 What ratio?

If a staff–pupil ratio in a school is 1 to 12, and a new teacher is employed, what is the new ratio? What if the new teacher is assigned to a new intake group of 20 pupils? Is the ratio likely to go up or down overall?

Comment

You cannot, of course, work it out, because it depends on how many staff (and hence how many pupils) there are to begin with. If there were only 12 pupils to start, and 1 teacher, then the extra teacher would change the ratio to $2:12$ or $1:6$, a dramatic change. If there were 120 pupils (and hence 10 teachers) the extra teacher would change the ratio to $11:120$, which is hardly different. If the school were even larger, the overall effect would be even less.

If 20 new pupils are admitted, then since the new teacher is 'matched' to more than the average of 12, the overall ratio will decrease (move from $1:12$ to $1:$ something larger than 12, depending on the size of the school).

Task 19 Review

In your own words state:

- the difference between a ratio and a proportion;

- how a ratio or a proportion can be simplified;

- how to solve questions of the following type: you are told that a specified proportion of things or people have a certain property, and you are told how many things or people there are. How many have the property?

- how to solve questions of the following type: you are told that a specified proportion of things or people have a certain property, and how many actually have that property. How many things or people are involved altogether?

> **Strategies for avoiding common errors in using ratio and proportion**
>
> Be aware of whether you are dealing with a ratio or a proportion – ratio is a relationship between two (or more) quantities, e.g. 1:2; and proportion measures the ratio of one quantity to a whole, e.g. $\frac{1}{3}$ of something.

Calculating strategies

There is a common misconception that particular calculations need to be done mentally, or written down or done by using a calculator. In practice the division between methods is usually less clear-cut.

For example, the legal requirement to maintain a register of pupil attendance involves:

- twice-daily marking of those pupils present and those absent (with letters inside the circle to indicate type of absence (no letter indicating unauthorised absence));
- total pupils present for each session;
- weekly, termly and annual summaries for each individual and the class (these are often done by a clerical assistant).

> **Task 20 Doing the register**
>
> Complete the register extract in Figure 1. Note *how* you do each calculation.

TERM ENDING ....

	WEEKLY ATTENDANCE																	TERM TOTAL	AUTHORISED	UNAUTHORISED	TERM TOTAL	NUMBER
	1	2	3	4	5	6	7	8	9	10	11	12	13	14	15	16	17					
	8	10	10	10	10	8	10		10	10	9	10	10	8						0		1
	8	10	8	6	10	10	10		0	0	0	0	0	0					0	0	0	2
	8	10	10	4	10	10	8		10	10	10	10	10	8						2		3
	0	10	10	10	7	8	10		10	10	10	10	10	8						0		4
INDIVIDUAL PUPILS	8	10	10	10	10	10	10		10	10	10	10	10	8						0		5
	8	9	10	10	8	10	10		10	10	0	0	10	8	End of term					0		6
	8	10	10	10	10	10	10		10	10	10	10	10	8						0		7
	8	10	10	9	10	10	10		10	10	10	10	10	8						0		8
	8	10	10	10	6	10	10		10	10	10	10	10	8						0		9
	8	10	10	10	10	10	10		0	10	10	10	10	8						0		10
	0	10	10	10	10	10	10		8	10	10	10	10	8						0		11
	8	10	10	10	10	10	10		10	10	10	10	10	8						0		12
	0	0	0	0	0	0	0		10	10	10	10	10	8					0	0	0	13
													TOTALS FOR THIS TERM								

TOTAL ATTENDANCES FOR THE TERM	
TOTAL TIMES OPEN FOR THE TERM	126
AVERAGE ON ROLL FOR THE TERM	12
AVERAGE ATTENDANCE FOR THE TERM	
TOTAL AUTHORISED ABSENCES FOR THE TERM	
TOTAL UNAUTHORISED ABSENCES FOR THE TERM	
PERCENTAGE UNAUTHORISED ABSENCES FOR THE TERM	

Figure 1 Register extract

Comment

An experienced school secretary was asked how she would do this task. She would:

1. Fill in the absences (noting that weeks 1 and 14 were only eight sessions).

2. Add up the attendance of the two pupils who were only on the roll for half a term (Pupils 2 and 13) by mentally counting the 10s and then the other numbers, jotting down the sub-totals.

3. For the rest of the class, mentally subtract absences from 126 (the total sessions open for the term).

4. Mentally add the absences columns, checking that total = authorised + unauthorised, and transfer to the bottom columns.

5. Use a calculator to add the Term Total column and check by redoing in the reverse direction, and transfer.

6. Use a calculator to work out the average attendance and percentage unauthorised.

She then added:

> I would then have to transfer the authorised and unauthorised totals to the main school record for the governors' report, prospectus and LEA returns… I never understood why it was thought necessary to work out individual class percentage unauthorised … I suppose it might highlight a problem class but if class sizes are very different the absence of one child has a disproportionate effect.

How you did each calculation is likely to have depended on such factors as:

- your confidence in manipulating numbers – particularly those arranged horizontally;
- how much practice you have had recently;
- your armoury of mental strategies;
- the availability of a calculator.

In this case calculating 'average attendance' required division by 12, which you may have done mentally, by a written method or using a calculator. In practice, a class size would not normally be as low as 12, so you would probably use a calculator.

The particular method you used for completing each section of the register extract will have depended on your own knowledge and confidence with the particular task – your choice of method may have been different faced with different numbers or calculations. In solving other numerical tasks your choices might be different again depending on such issues as:

- the time available;
- the technology available (calculator, computer);
- whether an exact answer is needed;
- how the result is to be used, and by whom.

Task 21 Theatre trip cost

Imagine you are planning an out of school (i.e. voluntary) theatre trip for Year 10 to see a production of their set GCSE Shakespeare play. You need to send out a letter giving the trip cost and inviting parents to sign up for their children.

You have the following information:

- there are 215 pupils in Year 10;
- seats cost £5.45;
- a 45-seater coach for the trip costs £75;
- school policy is to have one adult per ten pupils.

How do you work out what to tell parents about the cost of the trip? Think about what you do mentally, what you write down, and where you use a calculator – and why.

Comment

There are many ways of working out the estimate. Only an approximate answer is needed but there are dangers in approximating too soon.

1. A first rough estimate

$$£2 \text{ per seat in the coach } + £5 \text{ per ticket } = £7 \text{ per head}$$
$$\text{(overestimate)} \qquad \text{(underestimate)}$$

But will this cover every eventuality? Advertise at £8 to be 'safe'.

2. A more precise estimate: two methods

Method A: Assume you will have only full coaches. (This may mean that some pupils are turned away.)
One coach can take 40 pupils, which will mean taking four adults, so the total cost per coach is

$$£75 + (44 \times £5.45) = £314.80$$

Then the cost per pupil is £7.87 (assuming that adults go free). To allow for dropouts or changes in the coach charge, and to make collection easy, it would be best to advertise at £9.

Method B: Work from an estimate of the number of pupils likely to want to go. Say, based on past experience, that about half the year group will want to; about 110 (so divisible by 10) pupils. The total party will be 110 pupils and 11 adults.

For 121 people you need three coaches, so the total cost is

$$(3 \times £75) + (121 \times £5.45) = £884.45$$

so the cost per pupil is £8.04.

Again, plan to charge £9.

There are always realities that can sabotage such cost calculations: a 'worst case' needs to be worked out. What if only, say, 82 pupils actually sign up? Then you would need only nine adults but still need three coaches. Will £10 per pupil cover the costs? This kind of 'what if' thinking is easily transferred to calculations with the aid of a spreadsheet. A new costing gives

$$(3 \times £75) + (91 \times £5.45) = £720.95$$

The cost per pupil is now £8.79 so £9 would cover it. But the initial rough estimate of £7, even if increased to £8, would not suffice.

Calculating mentally

Mental strategies are often idiosyncratic: they depend on your own understanding of numbers, the actual numbers involved and how many numbers you can 'hold' in your head. This idea is difficult for people used to having a 'set' method for any particular calculation.

> **Task 22   Calculating mentally**
>
> For each of the calculations below do the calculation mentally and make a note of the answer. Make notes or jottings explaining how you arrived at the answer.
>
> 1. 36 + 28
> 2. 57 − 29
> 3. 17 × 5
> 4. 60 × 50

Comment

A group of teachers worked on the calculations given above and compared their methods.

1. Many of them carried out 36 + 28 by separating the numbers into tens and units and then doing 30 + 20 then 6 + 8 and adding the two answers together. When working mentally many of us deal with the bigger numbers (in this case the tens) first, although when carrying out standard column addition or subtraction we are often told 'you have to start with the units'.

2. For 57 − 29 a popular method was to work out 57 − 30 then add 1. This method, sometimes called 'adjusting' or 'rounding and compensating', depends on spotting the fact that one of the numbers is close to another number, which is easier to calculate with. This means that the method used for these numbers may not be the method you would choose to use for another two-digit subtraction. Mental calculation is thus different from standard methods of written calculation in that the method chosen may well depend on the numbers.

3. For 17 × 5 there were two main methods. The first was to separate the 17 into 10 + 7 then multiply each of these by 5, leading to (10 × 5) + (7 × 5) = 85. This method is similar to that underlying formal long multiplication, although when carried out mentally tens are often dealt with before units. Another method used depended on the fact that 5 is half of 10. So 17 × 5 = half of 17 × 10 = $\frac{170}{2}$ = 85.

4. For 60 × 50 the most popular method depended on knowing that 6 × 5 = 30. Explaining how this leads to 60 × 50 = 3000 proved harder, with many just remembering that they'd been told to 'add noughts'. Another explanation was that 60 × 50 is 6 × 5 × 10 × 10. Another method for this calculation was to multiply by 100 then find half.

Calculating mentally with percentages

When working with simple fractions and percentages it is often quicker to use mental methods. These are made easier by knowing the fractional equivalents for common percentages. These include knowing such equivalences as 50% = $\frac{1}{2}$. It is worth remembering the conversions for commonly used fractions, decimals and percentages (Table 4).

Table 4 Commonly used fractions, decimals and percentages

Fraction	Decimal	Percentage
$\frac{1}{100}$	0.01	1%
$\frac{1}{20}$	0.05	5%
$\frac{1}{10}$	0.1	10%
$\frac{1}{8}$	0.125	12.5%
$\frac{1}{5}$	0.2	20%
$\frac{1}{4}$	0.25	25%
$\frac{1}{3}$	0.33	33.3%
$\frac{1}{2}$	0.5	50%
$\frac{2}{3}$	0.67 (to 2 d.p.)	67%
$\frac{3}{4}$	0.75	75%
$\frac{4}{5}$	0.8	80%

Many percentages can be worked out from other easier ones, as the examples in Table 5 show.

Table 5 Example percentage calculations

Calculation	Answer	Method
10% of 134	13.4	10% is $\frac{1}{10}$, so divide 134 by 10
20% of 134	26.8	Double 10%
25% of 134	33.5	25% is $\frac{1}{4}$, so divide 134 by 4. Alternatively, use 25% = 20% + 5% (i.e. twice 10% + half 10%)
12.5% of 134	16.75	12.5% is $\frac{1}{8}$, so divide by 8
15% of 134	20.1	15% is 10% + half of 10%
17.5% of 134	23.45	17.5% is 10% + 5% + 2.5%: find 10%, add half of that, and then add half of the half.

The final calculation in Table 5 is useful because 17.5% is the VAT rate.

Mental calculations in context

When calculations arise as part of life or work, most people use mental methods as a first resort. Sometimes such calculations are carried out exactly but sometimes an approximate answer is enough.

> **Task 23 Using mental methods**
>
> Think back over the past few days and consider whether you have carried out any mental calculations as part of your work or otherwise. Make a note of the context and the mathematics involved.

Comment

Most of us carry out more mental calculations than we realise as part of our daily life. Many of these can be very straightforward, for example dividing a class of children into groups for PE or calculating how many sheets of paper or paintbrushes are needed. Calculations also often arise in relation to measurements of various types. One group of adults, who had made notes of the calculations they carried out, found that calculations involving time and money dominated.

> **Task 24   Calculating in context**
>
> Teachers carried out the following calculations mentally as part of their work. Carry out the calculations and make a note of your method.
>
> 1. As part of costing a school trip, a teacher calculated the cost of museum entry for 58 children at £2.50 each.
>
> 2. A PE teacher needed to calculate the finishing time of a football match starting at 4pm, allowing 45 minutes for each half of the match and a 10-minute break at half time.
>
> 3. A teacher is aware of the expectation that 75% of pupils reach Level 4 or above in SATs. She calculates how many children this means in her class of 32.

Comment

Several methods of calculating these are given:

1. (a)   It will cost £10 for every four children.

    $58 = (14 \times 4) + 2$

    So it costs £140 for the 14 sets of four children and £5 for the remaining two children.

    The total is £140 + £5 = £145.

    (b)   It will cost £10 for every four children, so it costs £150 for 60 children. Two fewer than 60 children are going, so subtract 2 × £2.50.

    £150 − £5 = £145

(c) Split the cost into £2 + 50p.

   58 × £2 =   £116

   58 × 50p =   £29

   £116 + £29 = £145

(d) Calculate 50 × £2.50 then 8 × £2.50.

   100 × £2.50 = £250 so 50 × £2.50 is half of £250, which is £125.

   8 × £2.50 can be calculated by doubling £2.50 three times, to give £20.

   The total cost is £125 + £20 = £145.

2. (a) Find the total time taken by both halves and the interval and then count on this far from the starting time.

   Two halves lasting 45 minutes take 1 hour 30 minutes altogether.

   Adding 10 minutes for the interval gives 1 hour 40 minutes.

   Counting on 1 hour 40 minutes from 4pm gives a finishing time of 5.40pm.

   (b) Count on from 4pm to the end of the first half, then to the end of the interval, and then to the end of the second half.

   The first half ends at 4.00 + 45 minutes = 4.45pm.

   The interval ends at 4.45 + 10 minutes = 4.55pm.

   The match ends at 4.55 + 45 minutes = 5.40pm.

There are many other methods that include combinations of these two. Such methods make use of facts that many of us already know about time, such as 'double 45 minutes is an hour and a half' or '4.55 is five minutes before 5 o'clock'. Many football players or fans also already know the total time for a match including the interval. They also know that the match is unlikely to end at exactly the predicted time, due to injuries and other stoppages.

3. (a) 75% = 50% + 25%

   50% of 32 is 16.

   25% of 32 is 8.

   Therefore 75% of 32 is 24.

   (b) 75% is $\frac{3}{4}$.

   $\frac{1}{4}$ of 32 is 8.

   $\frac{3}{4}$ of 32 is 24.

Both methods are made easier by knowing the fractional equivalents for common percentages. The first method relies on knowing that 50% is $\frac{1}{2}$ and 25% is $\frac{1}{4}$, while the second method converts the percentage to a fraction immediately. Some teachers might simply work out 25% or $\frac{1}{4}$ of the class to see how many might *not* be expected to reach Level 4. Unless the number in the class is a multiple of 4, the answer will not be a whole number, so rounding will be required.

Written methods

In the context of school mathematics, 'written calculations' is usually taken to mean formal methods; however, calculations are often written down informally.

Task 25 Why write?

Think of some recent occasions when you have written down a calculation that you carried out either as part of your work or for some other reason. Try to decide why the calculation was written down.

Comment

We write down calculations for a variety of reasons. The fact that something is written down does not necessarily mean a written method of calculation has been used; it may be that the numbers are written down simply so that they are not forgotten and then mental methods or a calculator are used. Some examples of occasions when calculations are written and the possible reasons are given below.

- A teacher is carrying out preliminary costing for a school trip. He works out the cost for each possible part of the trip using a calculator. The calculations are written down in order to share them with colleagues when deciding whether to go ahead with the trip. If the trip goes ahead the calculations will be needed as a record of how the money was spent.
- A technology teacher is working out how much stock to order for the coming term. She works out in her head what will be needed for each project, then finds totals for the term. She writes things down in order to keep track of her early answers before using them again to find the total.
- A head of year is collating subject choices for eight classes and finding the total in the year group wanting to study each subject. She writes the numbers to be added underneath each other so that she can use the standard method of column addition to add them up.

Written methods of calculation are used when:
- calculations are too complex to be done entirely mentally, and a calculator is not immediately available;
- the process of calculation needs to be communicated to someone else (or yourself at a later date).

Calculators

Three main types of calculator are available:
- A basic calculator or four-function calculator. This is the type usually used in primary schools and is adequate for most calculations. In addition to the four functions (+, −, ×, ÷) it may also have some other keys, for example %. Some also have brackets and memory keys.
- A scientific calculator. This is the type usually used by pupils in secondary schools. It has more keys than a basic calculator, including trigonometric functions and memory keys, and sometimes statistical functions.
- A graphics calculator. This has a large screen that displays each key press. They are scientific calculators but also have graph drawing functions and special statistical features, and are commonly used around GCE A level.

Even within these three main categories there is variation between different brands and models of calculator. It is worth spending some time getting to know how your own calculator works. It can also be interesting to compare the way two different calculators, perhaps a basic calculator and a scientific calculator, deal with the same calculations.

The next task draws your attention to some of the features of a calculator operation. The intention is to raise issues through your being surprised at the answers the calculator gives.

> **Task 26 Guess and press**
>
> For each key sequence below, guess what the final display will be then key in the sequence and see if you were right. If you have access to both a basic calculator and a scientific calculator, try the exercise with both and note any differences.
>
> 1. **2 [+] 3 [+] [=] [=]**
>
> 2. **2 [+] 3 [×] 4 [=]**
>
> 3. **2 [÷] 3 [=]**
>
> 4. **2 [÷] 3 [=] [×] 6 [=]**
>
> 5. **9 9 9 9 9 9 9** (press more 9s if your calculator has space) **[+] 1 [=]**

Comment
Surprising as it may seem, the answers will vary depending on the calculator you used. Some of the possible answers and the issues they raise are discussed here.

1. There are no prizes for guessing that pressing 2 [+] 3 [=] will give the answer 5 on any calculator. On some calculators pressing [=] again will have no effect, but on some the calculator will add 3 again, and will keep adding 3 if you keep pressing [=]. Others add the 2 again and again. This is because of the calculator's constant function, which is a way of repeating the same operation. On some calculators this is activated by pressing [+] twice rather than [=] twice. On some calculators a k appears in the display to show that the constant function has been activated.

2. Some calculators will give the answer 14 to this calculation, while some will give the answer 20. This will depend on the order in which the calculator does the operations, which in turn depends on the type of logic it uses. Calculators using an algebraic logic (and hence following the rules of algebra and carrying out the multiplication before the addition) will do $3 \times 4 = 12$ first, and then add 2, giving the answer 14. Most scientific calculators have this type of logic. Calculators with arithmetic logic, on the other hand, will carry out the operations in the order the keys are pressed. Thus $2 + 3 = 5$ will be done first, then $\times 4$, giving the answer 20. Most basic calculators have this type of logic.

3. The calculation $2 \div 3$ is problematic for the calculator as the answer, 0.6 recurring ($0.\dot{6}$) in theory 'goes on for ever'. The calculator can only show as many digits as it has space for in the display. Some calculators will simply 'cut' the answer off, showing as many 6s at it has space for. Others will round the answer up so that the final digit shown is 7.

4. This calculation is a continuation of the previous one. Some calculators will treat the expression as a whole and give the correct answer of 4. Others will take the cut off or rounded answer to the first part of the calculation and multiply that by 6, hence giving a slightly incorrect answer (e.g. 3.999 999 9) due to rounding error.

5. Filling the display with 9s has displayed the largest number the calculator can show and thus trying to add 1 may lead to an error. Usually an E will appear in the display to indicate that a calculation cannot be done, although an incorrect answer often appears as well, so it is important to take note of the E. Some calculators simply write the word 'error' in the display. There are calculators that can carry out this calculation and they do so by writing the answer in standard form. So if you have pressed **9 9 9 9 9 9 9** [+] **1** [=] the calculator may reach the correct answer, 100 000 000 or 10^8, which is written in what is known as standard form and may appear on the calculator as 1. ^{08} meaning 1.0 to the power 8 (if your calculator had space for more than seven 9s, then the result will be 1.^{n} when n is a number greater than the number of 9s entered) If your calculator automatically goes into standard form when the numbers involved become very big or very small then it is useful to be aware of this so you know what is happening, even if you do not wish to make use of the standard form facility.

Money calculations

Calculators are commonly used to carry out calculations involving money, especially when several large amounts are involved and an exact answer is needed.

> **Task 27 Shopping list**
>
> A food technology teacher purchases a range of items at varying prices:
>
> £3.50   £1.65   95p   £2   9p
>
> Use a calculator to find out the total cost of these items, and think about what you needed to do when entering the prices.

Comment

Although this is a fairly straightforward calculation, the main point to watch out for is that you needed to work in either pounds or pence and be consistent. If you work in pounds then 95p should be entered as 0.95 and 9p as 0.09. Adding these to the amounts already expressed in pounds will give an answer of £8.19. If you work in pence then £3.50 should be entered as 350, £1.65 as 165 and £2 as 200. This will give an answer of 819 pence, which should be converted to pounds to give £8.19.

> **Task 28 School visit**
>
> A teacher is taking 48 children to the zoo, where the entrance fee is £5.35. Use a calculator to find the cost for all the children. (Assume the teacher is unable to negotiate a group discount!)
> The teacher also has to pay for four adults at £6.95 each. Find the total cost for the party.

Comment

The first calculation is straightforward, although there is a possible difficulty in interpreting the answer. If you have worked in pounds and entered 5.35 × 48 the calculator display should show 256.8, and this needs to be written as £256.80.

The second part of the question raises the issue of how you carry out a calculation involving several steps using a calculator. Here are some possible approaches for this calculation:

- Jot down the answers to the two multiplications £5.35 × 48 = 256.80 and £6.95 × 4 = £27.80, then add them with the calculator giving a total of £284.60.

- In this case the calculation can be done without writing anything down if you have a calculator with algebraic logic, by pressing

    5.35 [×] 48 [+] 6.95 [×] [=]

- If you have a calculator with arithmetic logic and brackets you can press

    (5.35 [×] 48) [+] (6.95 [×] 4) [=]

However, care does have to be taken when mixing operations on a calculator and if in doubt it can be better to write parts of the calculation down, as suggested above, or use the memory facility of the calculator, as suggested below.

The memory facility of the calculator can be used to store the answer to the first calculation by pressing [Min] when the result of the first calculation is shown. The calculator can then be cleared and the second operation carried out, and then the first added to it, recalling it by pressing [MR]. The memory keys on calculators vary. For example, some have [M+], which adds the number in the display to the memory. If you are using the calculator memory then it is worth checking that the memory is empty (usually by pressing [MC] or by putting 0 in the display then pressing [Min]) before you start.

Calculators and percentages
When carrying out calculations involving percentages using a calculator, there is no need for a % key, although some calculators do have one. It is important, however, to keep in mind that percentage means 'out of a hundred' and that 'of' can be thought of as multiplication.

If, for example, you wish to find 15% of 480 using a calculator you could press

$$15 \; [\div] \; 100 \; [\times] \; 480 \; [=] \; 72$$

A related method is to convert percentages to their decimal equivalents before calculating, so 15% becomes 0.15.

Some final comments on calculators
Calculators will be most people's first choice for carrying out calculations, especially with complicated numbers or operations. For simple calculations, however, it can be quicker to work mentally.

Strategies to avoid common errors using calculators

- Remember to press the clear key before starting a new calculation.

- When you have entered a number check by looking at the display that the number entered is correct.

- Mentally estimate the answer and check that the calculator solution is of the right order (about the right size).

- Take account of how the calculator being used performs multiple operations. In particular, make sure that operations are entered in the correct order, especially when carrying out calculations involving multiplication/division and addition/subtraction.

- Make sure that the answer makes sense in the context of the question. (For example money answers should be rounded to the nearest pence, and answers about numbers of people should be whole numbers!)

Spreadsheets

Spreadsheets, which are included in any computer office package (e.g. Microsoft Works, AppleWorks, Microsoft Office), are relatively easy to learn how to use at a sufficient level to take the tedium out of many school administrative tasks involving calculations, such as school trip cost estimates and accounts, resource orders, set lists and class or equipment checklists.

Being able to use a spreadsheet is now an ITT (Teacher Training Agency) requirement. Spreadsheets can be very powerful tools for analysing pupil performance (see pp. 59–60).

Getting started
When you load up a spreadsheet you will see a grid of **rows** and **columns** forming a series of boxes (**cells**). Each cell has its own identity or address, for

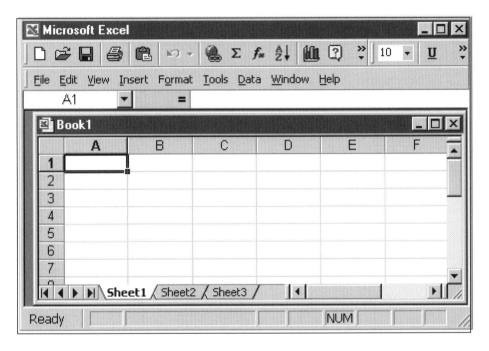

Figure 2   Spreadsheet showing cell A1 highlighted

example the top left-hand box (shown with a black border) is A1 i.e. column A row 1 (Figure 2). A cell may contain *text*, a *number* (in various forms) or a *formula* (or nothing!). For example a simple resource order might appear as in Figure 3.

Once a spreadsheet is set up and saved, changes can be made so it can be used again and again. For example, in the one shown in Figure 3, changing the equipment and quantity required will automatically recalculate the total cost of the order (so quantities can be adjusted to match the budget available!).

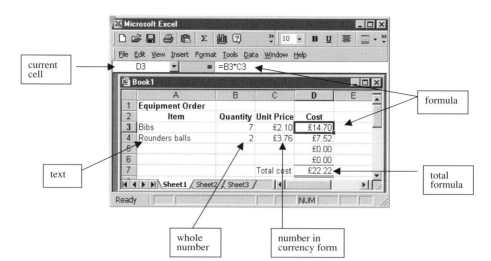

Figure 3 Spreadsheet showing a simple resource order

For more on using spreadsheets see the section on formulas, (p. 59).

CHAPTER 3

Measures and Formulas

The nature of measures and measurement

Measurements are made in particular contexts and for particular purposes. These determine both the units of measure and the quality of measurement required.

> **Task 29 How big?**
> Think of a school you know. How big is it?
> The task is not to answer that question, but rather to think of situations where the question might be asked and thus what measurements might be needed.

Comment
The usual response to such a question is an approximate number of pupils. However, there are a number of reasons why other measures of school size may be needed. Each has a different purpose, which determines the measure to be used, the quality of measurement required and the units used:

- The planned pupil intake number is used to decide the number of places available at the school, which in turn can influence the facilities to which the school is entitled.

- The count of pupils on the roll to the nearest five pupils is used to decide the pupil-related element of the budget allocation.

- The distance around a school site to the nearest metre would be needed to cost the provision of perimeter fencing.

- The area of the grounds is used to decide the cost of the maintenance contract.
- The number and sizes of classrooms, halls, corridors, toilets, etc. is used to decide the cost of the cleaning contract.
- The volume of the school buildings in cubic metres is used to decide the boiler capacity required for central heating.

The purposes of a measurement determine the quality of measurement required (and therefore the tool or instrument used to measure), and what is included and what is ignored.

Task 30 Teaching time

Comment on the following:

In a secondary school each lesson period is 40 minutes. There is therefore twice as much teaching time in a double period as in a single.

Comment

The lesson period includes the time needed for pupils and staff to change classrooms (pack bags, move, unpack bags). So the start of teaching time could be 5 minutes or more after the start of the period. This means that for a single period the teaching time is approximately 35 minutes; but for a double period it is 75 minutes. (For practical lessons the teaching time would be even less because of the need for getting out and putting away equipment and, for PE, changing time.)

When comparing data that arise from measuring, it is essential that like is being compared with like. For example, the Third International Mathematics and Science Survey (TIMSS) compared test results of a particular year group across various countries. In many countries pupils are not allowed to advance a year group unless they have achieved a particular standard; in the UK pupils normally advance by age. Not surprisingly the UK appeared to do badly in some of the comparisons.

Using measurements in calculations

When a measure is expressed in different units the relationship between numbers is exact: 2170 mm is identical to 217 cm. However, actual measurement using instruments is always approximate: it is not possible to say that the width of a room measures exactly 2170 mm. This is partly due to human error in, for example, lining up the measuring mark exactly opposite the object being measured; but is also due to the nature of measuring and the fact that no measuring instrument is without some inherent error. Even when you had measured as carefully as you could, if you were able to magnify around the measuring mark there would still be further variation from your mark. Craftsmen allow for this by including leeway; they tend to cut 'too large' and then trim to fit. The same principle is used when doing budget forecasts to avoid overspend.

Even digital electronic measuring devices have an inherent error. For example, when an electronic weighing scale that is able to measure to the nearest milligram produces a reading of, say, 34 mg, then the object it measured might be slightly above or slightly below 34 mg (33.5 mg–34.4 mg to 1 d.p.). The display is unable to distinguish between those weights. Imagine weighing some rice and either adding or taking off a few grains. There would need to be considerable change before the display registered a different figure.

It is important to appreciate that any errors will affect the result of any subsequent calculation. For example, if a bookcase is quoted as having a width of 84 cm this implies 84 cm *to the nearest centimetre*, so there could be an error of half a centimetre (5 mm) on either side of 84 cm. All that can be said about the actual measurement is that it lies somewhere between 83.5 cm and 84.5 cm. Where an error or tolerance is quoted as a percentage it is called the **relative error**.

> ### Task 31   Library problem
> It is planned to use a run of ten bookcases each 84 cm wide. How much space must be allowed to ensure they fit?

Comment
The simple answer is 8.4 m (because 10 × 84 cm = 840 cm = 8.4 m).

But allowing a possible error of 0.5 cm per bookcase gives an error range of (10 × 0.5 cm = 5 cm) each side of 8.4 m. So, using the maximum, the space needed is 845 cm.

Of course, the actual space will also have to be measured, and with its own error!

Units

The importance of standard measurements becomes evident when you want to communicate a measurement to someone else. In the UK there are three systems of units of measurement in use: imperial, metric and SI (Système International). When imperial measures were first used they were based on the sorts of amounts that people commonly used, or on parts of the human body. This means the relationships between units are unplanned and seem very arbitrary; for example, the length of a cricket pitch is 22 yards, which is equivalent to the old measure of one chain (an acre is 10 square chains). By contrast, the metric system was designed as a whole with the relationship between units following a logical pattern. SI units are a version of metric units constructed to make international communication simpler, especially in science and engineering.

The metric system

Metric measures are designed to fit together by being based on multiples of ten. For example, the distance between towns may be measured in kilometres (abbreviated to km); the length of a soccer pitch in metres (m); and the size of a whiteboard in centimetres (cm) or millimetres (mm). The prefixes 'kilo', 'centi' and 'milli' are used throughout the metric system, and there are other prefixes that are much less commonly used (Table 6).

These simple relationships mean that it is easy to convert measurements in, say, centimetres to millimetres or metres; for example, 2170 mm = 217 cm = 2.17 m. For measurements of furniture millimetres are usually used rather than centimetres as this avoids the need to use a decimal point.

Table 6   Metric prefixes

Prefix	Symbol	Meaning	Example
giga	G	thousand million	gigabytes
mega	M	million	megaton
kilo	k	thousand	kilowatts
hect(o)	h	hundred	hectare
deca	da	ten	decade
deci	d	tenth	decibel
centi	c	hundredth	centilitre
milli	m	thousandth	milligram
micro	µ	millionth	micrometer (micron)
nano	n	thousand millionth	nanosecond

Converting units

It is common to need to convert measures within a system of units such as imperial or metric, and between different systems of units. With the exception of time, it is becoming increasingly rare to need to convert within non-metric systems.

> **Task 32   Theatre trip timing**
>
> It is planned to take a group of pupils to a theatre performance starting at 7.30pm and finishing at 9.45pm. The coach company says it will take 1 hour 45 minutes to get from the school to the theatre. What time should the pupils be asked to meet, and what time should parents be told to pick up the pupils?

Comment

A variety of notations are used to separate hours and minutes. Examples are 7-30, 7.30 and, increasingly, 07:30 (the way it is often shown on digital clocks). The task seems, at first glance, to be a question involving the subtraction and addition of 1 hour 45 minutes. You probably did that by treating it as 2 hours and then adjusting by a quarter of an hour rather than doing formal calculations involving hours and minutes.

But if the task is a 'real' question then it becomes more complicated, involving estimates of loading and unloading times, making allowances for the unexpected on the journey, and so forth. In reality you would probably over-estimate for the outward journey and under-estimate for the homeward one.

Even within a single system of units it can be confusing to convert one unit to another, especially in units such as those for area.

> **Task 33 Square units**
> The area of a piece of carpet is 1.5 m². What is this in cm²?

Comment
A common mistake is to say that because there are 100 centimetres in a metre,

$1.5 \text{ m}^2 = 1.5 \times 100 = 150 \text{ cm}^2$

This is not the correct answer. If you need convincing, draw a square and mark the sides 1 m = 100 cm.

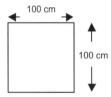

The area of the square is 1 m × 1 m = 1 m² or 100 cm × 100 cm = 10 000 cm².

So an area of 1.5 m² equals 1.5 × 10 000 = 15 000 cm².

Converting between imperial and metric units

Because both metric and imperial units are still used in the UK, people frequently have problems when measures are expressed in a system different to the one they normally use. Most people find that they are confident with only one system of units for any particular purpose. This may be metric for some measures and imperial for others.

They may, for example, feel happy measuring furniture in millimetres or buying petrol in litres, but use miles for distances. Many people refer to a 1 kg bag of sugar as 'a two-pound' bag although it is actually 2.2 lb. Where there is a need in everyday use to convert from one system to another, most people use easy reference points such as:

- 30 cm is about the same length as a foot;
- 1 m is slightly longer than a yard;
- 5 miles is about the same as 8 km.

For most everyday purposes, these are usually good enough. For more exact conversions you need to use a calculation method using a conversion (or scale) factor. If you have several conversions to make, it may be quicker to use the constant facility on your calculator. If you needed to do a number of conversions you could produce or use a conversion table or a conversion graph. All these methods require an equivalence point other than zero from one system to the other.

Conversion formulas and tables are often included in diaries:

To convert miles to kilometres multiply by 1.609. To convert kilometres to miles multiply by 0.6214.

Table 7 Miles–kilometres conversion table

Miles		Kilometres
0.621	1	1.609
1.234	2	3.219
1.864	3	4.828
2.468	4	6.437
3.107	5	8.047
3.728	6	9.656
4.350	7	11.265
4.971	8	12.875
5.593	9	14.484

Task 34 Using a conversion table

Convert the following distances using Table 7:

1. 50 miles to kilometres
2. 95 km to miles

Then calculate them by applying the formulas: 'To convert miles to kilometres multiply by 1.609: to convert kilometres to miles multiply by 0.6214'.

Comment
From Table 7:

1. 5 miles is 8.047 km, so 50 miles is ten times that, i.e. 80.47 km.
2. 90 km is 55.93 miles and 5 km is 3.107 miles so 95 km = 55.93 + 3.107 = 59.037 miles.

Using the formulas:
1. 50 × 1.609 = 80.45 km

2. 95 × 0.6214 = 59.033 miles

The answers obtained are not exactly the same due to rounding errors. For this reason it is usual to quote to the nearest unit. This would then give the following solutions:

1. 50 miles = 80 km to the nearest kilometre.

2. 95 km = 59 miles to the nearest mile.

Once a formula or conversion factor is known a conversion graph can be produced either by plotting two known points ((0,0) can be a useful one) and drawing the straight line passing through both, or by using a spreadsheet program. Note that (0,0) cannot be used for interval scales when the zero is arbitrary e.g. 0°C is not equivalent to 0°F. A graph can be an easier presentation to use in particular circumstances, for example converting kilometres to miles for use on a school trip abroad (Figure 4).

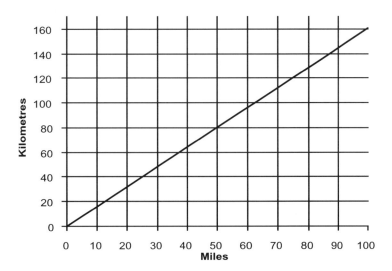

Figure 4 Conversion graph for miles–kilometres

Task 35 How far?

From the graph in Figure 4, read off the conversion for the following distances:

1. 80 km in miles
2. 50 km in miles
3. 70 miles in kilometres
4. 60 miles in kilometres

Comment
1. 50 miles
2. 30 miles
3. 115 km
4. 95 km

Note the same graph could be used to convert miles per hour (mph) to and from kilometres per hour (km/h, or km h^{-1}) because both speeds are 'per hour'.

Task 36 Converting currency

You are preparing for a school trip to Europe. Produce a currency conversion table and conversion graph for use by yourself and the pupils. Either use the current exchange rate for Euros (€) or a particular country's currency (available from newspapers, banks, post offices, etc.) or use 1 € = £0.58.

Comment
A conversion rate of 1€ = £0.58 means to convert Euros to pounds multiply by 0.58. 'Undoing' this says to convert pounds to Euros divide by 0.58 (or multiply by the reciprocal of 0.58, i.e. $\frac{1}{0.58}$ = 1.724 (correct to 3 d.p.)).

Table 8   Pounds – Euros conversion table

£		€
0.00	0	0.00
0.58	1	1.72
1.16	2	3.45
1.74	3	5.17
2.32	4	6.90
2.90	5	8.62
3.48	6	10.34
4.06	7	12.07
4.64	8	13.79
5.22	9	15.52

To produce a conversion graph, plot two points (two known corresponding values) and draw a line through them.

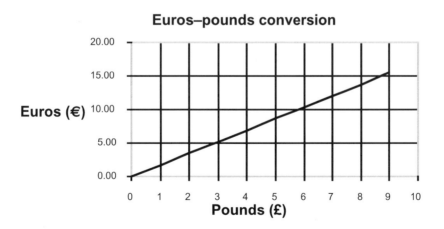

> **Strategies for avoiding common errors using measurements**
>
> - Make sure that mixed units are not inappropriately combined in calculations (perhaps by retaining units when working with quantities).
> - No rounding should be done until the end of any calculation.
> - Where quantities involved in a calculation are not all known to the same number of significant figures, the result can only be relied on to one significant figure *less* than the quantity that is known least exactly.

Using formulas

Many people are not confident in using formulas in mathematical notation. This is usually because a sense of panic can be engendered by a sea of letters. If you are confident with algebraic manipulation, formulas are really just a particular kind of equation and so can be used and manipulated in the same way as equations. Formulas relate to some specific relationship; they are not just letters and numbers. In one sense formulas are a set of shorthand instructions for calculating. Formulas normally refer to quantities and so care has to be taken with the units of measure, particularly compound measures, e.g. mph.

Translating formulas

> **Task 37 Shorthand**
>
> The following sentence is taken from a diary:
>
> To convert miles to kilometres multiply by 1.609.
>
> Try to write this calculating instruction as succinctly as you can.

Comment
Miles times 1.609 equals kilometres.

Or perhaps even more succinctly, miles × 1.609 = km.

This can be made even more concise by making *m* stand for distance in miles and *k* for distance in kilometres:

$1.609m = k$

A word instruction can be written in shorthand as an algebraic formula; both have the same meaning.

One way to overcome anxiety in dealing with algebraic formulas is to translate them into words so the meaning becomes clear to you.

> **Task 38 From letters to words**
>
> In the following formula, *k* is distance in kilometres and *m* is distance in miles. What does it mean?
>
> $0.6214k = m$

Comment
kilometres × 0.6214 = miles

To convert kilometres to miles multiply by 0.6214.

Or even more fully, multiply a distance in kilometres by 0.6214 to get the corresponding distance in miles.

This process of translating to and from words and letters can be used for any formula. The formula for converting temperatures in degrees Fahrenheit (°F) to degrees Celsius (°C) is

$C = \frac{9}{5}(F - 32)$

In words this says:
1. Take the Fahrenheit temperature and subtract 32.
2. Multiply the result by 5 and divide by 9.

Remember that the order of calculation is important – brackets are calculated first. While the written instructions may seem easier to follow, the use of symbols is succinct and can make noticing generalities easier.

Task 39 Decoding a formula

The formula for converting temperatures in degrees Celsius to degrees Fahrenheit is

$$F = \tfrac{9}{5} C + 32$$

Write out the instructions in words.

Comment

Multiply the Celsius temperature by $\tfrac{9}{5}$ (or multiply by 9 and divide by 5).

Add 32. Remember that multiplication is done before addition.

Task 40 Changing temperature

Convert the following temperatures to the nearest degree in the other temperature scale:

1. 32°F

2. 60°F

3. 98.6°F

4. 100°C

5. 30°C

Comment
1. 0°C (freezing point of water)
2. 16°C (minimum office working temperature)
3. 37°C (normal human body temperature)
4. 212°F (boiling point of water)
5. 86°F (a hot day)

Using formulas in spreadsheets

A spreadsheet is an ideal tool for situations where a formula calculation has to be repeated many times, for example, when calculating weighted examination grades or doing budget predictions. The only snag is that of course the formula has to be entered correctly and that means deciding on the formula to use in the first place.

> **Task 41   Examination Percentages**
>
> A particular subject end of year examination has three parts:
> Part 1:   an oral marked out of 15
> Part 2:   a practical marked out of 20
> Part 3:   a written paper with five questions each worth 20 marks and up to 5 marks for spelling and grammar.
>
> How would you work out the final percentage marks for the class? (See pp. 15, 32 for work on percentages.)

Comment

The straightforward answer is, for each pupil, add up their marks in the three parts, then divide by the maximum marks (15 + 20 + 105 = 140) and multiply by 100 to get the percentage.

$$\frac{\text{total marks scored}}{\text{maximum marks}} \times 100 = \text{final percentage mark}$$

Doing this for a whole class could be tedious so it would help to set up a spreadsheet (Figure 5).

Different spreadsheet packages have slightly different ways of entering formulas and fixed references. Use the 'Help' facility or manual to check. Most spreadsheets (including Excel) have a cell format that converts to percentages. Column F on Figure 5 shows the formula to enter if your spreadsheet does not have this facility (*100 means × 100).

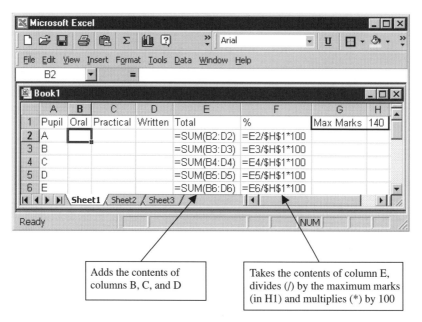

Figure 5 Spreadsheet calculating class marks as percentages

Once the formulas are set up for one row, they can be copied and pasted into the other rows. (Using $ makes the cell reference absolute, i.e. fixed, not relative to a row or column.) As the pupils' marks are entered, the spreadsheet will calculate each percentage mark.

The solution in Task 41 assumes that each mark has the same 'worth', whether it was achieved in the oral, the practical or the written test. However, this is not always the case.

Task 42 Weighted marks

A subject test has two parts:

T1: a multiple choice paper with 25 questions each worth 1 mark (total 25)

T2: a written paper with five questions each worth 25 marks (total 125)

If T1 is to contribute 10% of the final mark and T2 is to contribute 90%, how would you calculate the final marks for a class?

Comment
One way of doing this is to think of the first test contributing $\frac{1}{10}$ of the final mark and the second test $\frac{9}{10}$.

For the first part the contribution to the final mark is:

$$\frac{\text{pupil's T1 mark}}{\text{maximum T1 mark}} \times \tfrac{1}{10} \text{ of the final mark}$$

And for the second part:

$$\frac{\text{pupil's T2 mark}}{\text{maximum T2 mark}} \times \tfrac{9}{10} \text{ of the final mark}$$

The final mark is the sum of the two parts (of the final mark), which then needs to be turned into a percentage mark by multiplying by 100. (Each part could be turned to a percentage of the final mark by multiplying by 100, and then adding the percentages, since they are percentages of the same thing (the final mark).)

Example
A pupil receives 12 marks on T1 and 82 on T2. What is her final mark?

The contribution from the first part is:

$$\frac{\text{pupil's T1 mark}}{\text{maximum T1 mark}} \times \tfrac{1}{10} = \tfrac{12}{25} \times \tfrac{1}{10} = 0.048 \text{ of the final mark}$$

And the contribution from the second part is:

$$\frac{\text{pupil's T2 mark}}{\text{maximum T2 mark}} \times \tfrac{9}{10} = \tfrac{82}{125} \times \tfrac{9}{10} = 0.5904 \text{ of the final mark}$$

T1 is 0.048 of the final mark, which is 0.048 × 100% = 4.8%.

T2 is 0.5904 of the final mark, which is 0.5904 × 100% = 59.04%.

So the pupil's final mark is 59.04% + 4.8% = 63.84% = 64% to the nearest 1%.

A tempting alternative might be to argue that the marks are in the ratio 1:9 so the total weighted score is T1 + 9T2, and the pupil percentage mark can be calculated:

$$\frac{\text{pupil's weighted mark (T1+9T2)}}{\text{maximum weighted mark (T1+9T2)}} \times 100$$

However, this assumes that T1 and T2 correspond to fractions of the *same* number, which is not so in this case.

> **Task 43   It makes a difference**
>
> Convince yourself that using
>
> $$\frac{\text{pupil's weighted mark (T1+9T2)}}{\text{maximum weighted mark (T1+9T2)}} \times 100$$
>
> produces a different result than you would get using
>
> $$\frac{\text{pupil's T1 mark}}{\text{max T2 mark (25)}} \times \tfrac{1}{10} \times 100 + \frac{\text{pupil's T2 mark}}{\text{max T1 mark (125)}} \times \tfrac{9}{10} \times 100$$
>
> for the example in Task 42.

Comment
A good way to try out ideas is to pick some extreme marks (but not too extreme!) in each paper.

Applying different weightings spreads out the marks to differing degrees, and attaches differing values to each component.

Marking is easier when the allocation of marks matches the question format; weighting can then be applied to adjust relative values. Working out weighted marks is very time consuming if done 'by hand' rather than using a spreadsheet. Doing calculations like those in this section easily and automatically is exactly what spreadsheets were invented for!

Strategies to avoid common errors with formulas

- Translate formulas into words to ensure that they are meaningful, and so become aware if inappropriate solutions result.
- Formulas usually involve quantities, so make sure that units of measure are not mixed in calculations.
- Formulas involving fractions and percentages need care to ensure that the appropriate 'whole' is being used.
- The order of calculations must conform to algebraic logic: brackets → indices (powers) → division and multiplication → addition and subtraction. (Use the mnemonic BIDMAS to remember this order.)
- If in any doubt about the order in which a calculator or spreadsheet performs mixed calculations, use brackets to separate multiplication/division from addition/subtraction. Remember that basic four-function calculators use arithmetic and not algebraic logic.

CHAPTER 4

Statistics

Introduction

> **Task 44   What is 'statistics'?**
> The word statistics is used in a number of different ways. What do you understand it to mean? Jot down your immediate thoughts.

Comment
Depending on your previous experience you may have written something similar to one or more of the following. Statistics is:

- a subject or discipline, e.g. an A level mathematics module;
- methods of inquiry used to process numerical information;
- collections of data in general, e.g. the UK Government Annual Abstract of Statistics;
- particular numbers that somehow characterise a particular data set, e.g. averages.

The word can be used to mean all of these! Data collection and statistical thinking have a long history; the term 'statistics' derives from the German word *Statistik*, which was coined in the eighteenth century.

In this section we explore the second meaning: how numerical information, data, can be processed and to what uses the processed data can be put.

> ### Task 45 Some data
>
> Look at the following information and jot down your first thoughts:
>
> | | | | | |
> |---|---|---|---|---|
> | Alison | 58 | | Denny | 60 |
> | Peena | 36 | | Eva | 28 |
> | Curtis | 57 | | Fatimu | 52 |
>
> If you are then told that they are actually marks in a test, what thoughts do you now have?

Comment

'What are the numbers? What is it about?' They could be anything, weights, days absent, an identifier, marks out of 100, and so on.

However, if they are test results, 'How many questions were there?' might be one response and 'Why did Eva do so badly?' might be another. On further reflection you might ask questions such as 'How do the results compare with previous ones – for the group, for the individuals – or with other pupils of the same age, same sex, and so on?' And then you might ask the difficult question, 'Are these differences significant?' In other words, how large a difference would there need to be for you to decide that there was a real effect here and that differences were not simply due to chance variation?

A collection of data, numerical information, is of no value in itself; there needs to be a context. It has been said that statistics (in the 'a collection of data' sense) raise questions rather than provide answers! So there needs to be not only a context for collections of data but also a purpose: 'Why do I need to know this?' and 'How can the information be used?'

It is relatively easy to collect data and, using modern information and communications technology (ICT), relatively simple to process the data collected, but the analysis and interpretation are more problematic. Even more of a problem is how to respond to the processed information. Having the information is one thing: deciding whether it shows up significant differences and if the differences are significant what action, if any, can be taken to improve the situation is another.

Task 46 Who wants to know what?

Various groups of people have a legitimate interest in information about a particular school and its pupils. List some of these and jot down the main things that you think each group might wish to know.

Comment

Parents (prospective and current), pupils, teachers (applicants and current), governors, inspectors (local education authority (LEA) and Office for Standards in Education (OFSTED)) and researchers all have an interest in information about any particular school. They have questions to which they seek to get answers:

- Will my child be happy there?
- How well is my child doing?
- Is there a Sixth Form?
- What are the exam/test results like?
- How big are the classes?
- What is the behaviour like?
- How well is the school staffed and equipped?

Almost all the questions imply a comparison: how does the school/class/pupil compare with other local or similar schools/classes/pupils?

The process of obtaining the answers to such questions is cyclic. At the outset, specific questions are raised, and the relevant data are found or collected. The data then need to be processed and analysed, and the results interpreted. The interpretation may well raise new questions or require the refinement of the original question.

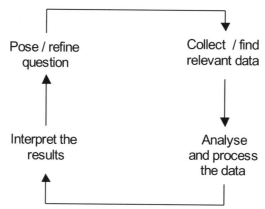

Figure 6 The cyclic nature of data collection and analysis

Individual class teachers, school managers, LEAs, government agencies and researchers collect a vast amount of data about such diverse matters as individual pupil performance, attendance, entitlement to free school meals, staffing and expenditure. Sources of such data include:

For an individual school:

- prospectus
- OFSTED report
- governors' report to parents
- Form 7 (biannual census of pupils on roll)
- reading test results
- verbal reasoning quotients

- baseline assessment
- teacher mark books
- registers
- Pre-Inspection Context and School Indicator Report (PICSI)

From LEAs and national bodies:

- national league tables
- LEA performance tables
- Autumn Package
- national summary results
- Performance and Assessment Report (PANDA)

- national benchmark information
- cost benchmarking
- comparators
- research reports and articles

Before you can make use of these data – whether to help improve practice in the classroom or for a whole school – they need to be analysed and interpreted. To do this it is necessary to:

- clarify the question under investigation;
- understand what information is needed and how it can be collected;
- know what statistical and graphical tools are available and use them correctly to describe situations, identify relationships and make comparisons;
- make inferences and draw conclusions about the question that was the starting point of the investigation.

Grouped data

Raw data can be processed in a variety of ways in order to make comparisons and to help make patterns and relationships easier to identify. A common process is to group the data. For example, in a particular school, test marks that measure performance on a reading test or a specific topic may be collected for each pupil. Looking at each pupil's mark on its own says little about the performance of the class overall, so the raw data can be grouped into the bands. In the case of reading ages, perhaps they could be grouped into month bands (9 years 0 months to 9 years 5 months, 9 years 6 months to 10 years 0 months and so on) or in the case of subject test marks they could be grouped in 10 point bands (0–9, 10–19, 20–29, …). By grouping the data in this way, the overall distributions of marks are more apparent – the teacher can see at a glance into which group most pupils' marks fall. Of course, in the process a great deal of detail has been lost and care has to be taken with any further processing.

The above example of test marks involved discrete data, but continuous data can also be grouped, for example, pupil heights or weights.

Some statistical terminology

Data (or facts and figures) involve quantities that can be measured on some kind of a scale. There are different categories of scale, based on the particular properties of what is being measured. You may already have met two such distinctions in the chapter titled 'Arithmetic'. These were firstly the fairly straightforward distinction between *names* and *numbers* and secondly the notion of *discrete* and *continuous* scales of measure.

> ### Task 47   Discrete and continuous
> If you haven't done so already, read 'Using and representing different types of numbers' (p. 7). Then spend a few minutes jotting down your understanding of the distinction between discrete and continuous scales of measure.

Comment
- **Discrete data** can take only a limited number of separate values; for example the year group in which a child is taught (e.g. Year 4, Year 5, etc.). Other examples of discrete data are the number of pupils on the roll and the number of half-day absences.
- **Continuous data** can have an infinite number of different values; for example within a given year group there are an infinite number of possible ages of each child: the number of possibilities is limited only by how accurately you choose to make the measurement. Other examples of continuous data are elapsed time and height.

Both names and numbers are frequently used in scales of measurement, and are especially seen on graphs. It is important to realise when numbers that involve numerical measurement are being used and when names (which are simply labels, but which may sometimes be numbers) are being used. There are two kinds of each type, which are described below.

Naming scales
- **Naming scales.** For example, a survey of the various subjects taught in a school would produce a naming scale consisting of mathematics, English, PE, etc.: a list of names (not numbers). Naming scales are simply names or labels and have no arithmetic properties, and no natural ordering.
- **Ordering scales.** Examples are days of the week or positions in league tables where schools are listed in order of achievement (first, second, etc.). Although an order is implied, and ordering scales are often numbers, there is no zero-point, nor is there a meaningful interval

between positions. A school that comes fourth may be considered to be better in some sense than one that comes eighth but cannot be said to be twice as good!

Numerical scales
- **Interval scales.** An example is the hours of the day. These are numerical and so there is an ordering (3 o'clock follows 2 o'clock) and also information on the intervals between points on the scale. However, as with ordering scales, the zero is arbitrary; for example the zero of the 24-hour clock does not imply 'no time'. Also, ratios have no meaning: 4 o'clock is not twice as late as 2 o'clock!
- **Ratio scales.** Examples are most standard measurements such as length, elapsed time (as opposed to time of day) and age. Unlike interval scales, these have the full range of properties of numbers. Zero really does mean a zero amount and the order of points, the interval between points, and ratios are meaningful. So a length of 0 cm really means zero and a length of 10 cm is twice as much as a length of 5 cm.

Data can also be categorised by source:

- **Primary data** refers to information that is collected for a specific purpose and may be largely unprocessed (unprocessed data are often called 'raw' data); for example, individual pupils' raw test marks, grades or percentages in a teacher's mark book.
- **Secondary data** refers to information that has already been collected by someone else for a different purpose but which can be used in a subsequent (second) investigation. Secondary data are usually at least partially processed and are more widely available than primary data; for example, Department for Education and Employment (DfEE) pupil performance data and, indeed, government statistics in general.

> **Task 48 What type of data?**
>
> You should now be familiar with the following four (not unconnected) ways of categorising data:
>
> - names, numbers;
>
> - discrete, continuous;
>
> - naming, ordering, interval, ratio;
>
> - primary, secondary.
>
> Use these four classifications to categorise the data below.
>
> 1. Marks in a school examination.
>
> 2. Family sizes of children in a particular class.
>
> 3. DfEE listing of schools in alphabetical order.

Comment
The types of data are summarised in Table 9.

Table 9 Categorisation of data in Task 48

	Numbers/ names	Discrete/ continuous	Ratio/ naming	Primary/ secondary
1. Exam marks	Numbers	Discrete	Ratio	Primary
2. Family size	Numbers	Discrete	Ratio	Primary
3. DfEE listing	Names	Discrete	Naming	Secondary

Statistical summaries

For data to be of use they need to be presented in forms such that patterns, relationships or summaries can be easily found. The common methods of summarising data include averages, measures of the spread and boxplot diagrams. These are each dealt with in the following pages.

Averages

Depending on the nature of the question under investigation, the type of average value used to represent the whole batch might be any of the following:

- the most common value (the **mode**);
- the middle value (**median**);
- a value calculated from all the data (**arithmetic mean**).

These are all averages, but when just the word average is used on its own, people tend to assume that what is being referred to is the mean. The mean is calculated by adding all the values in a batch of data and then dividing by the number of values in the batch. Such a calculation requires numerical values rather than names.

Pupil performance is categorised by National Curriculum attainment level for Key Stages 1–3 and external examination grade for Key Stage 4. For example, English Key Stage 1 performance has the following National Curriculum categories:

W   1   2C   2B   2A   3   4   5   6   7   8   EP

(W signifies working towards Level 1, EP is exceptional performance):

Whereas other subjects only have:

W   1   2   3   4   5   6   7   8   EP

So that pupil, school and national averages can be calculated and compared across subjects these categories are assigned numerical ranges. The mid-point of each range is used as a point score.

The National Curriculum attainment levels have the point scores shown in Table 10.

Table 10 National Curriculum attainment level point scores

Level	W	1	2C	2B	2A	3	4	5	6	7	8	EP
Mid-point	3	9	13	15*	17	21	27	33	39	45	51	57
Minimum of range	0	6	12*	14	16	18	24	30	36	42	48	54

* Values used for subjects other than English at Key Stage 1.

Task 49 Adding scores

A group of nine pupils achieved the following SATs results at Key Stage 2. Complete Table 11 and calculate the average for the group.

Table 11 SATs results at Key Stage 2

Pupil	English attainment level	Maths attainment level	Science attainment level	Point score*	Average	Average level
A	2B	3	2	15+21+15=51	17	2
B	4	4	4	81	27	
C	5	3	4			
D	2B	2	3			
E	5	5	5			
F	4	3	4			
G	4	4	5			
H	5	4	4			
I	4	4	4			

* See Table 10 for point scores.

Comment

Table 12 Completed table of SATs results at Key Stage 2

Pupil	English attainment level	Maths attainment level	Science attainment level	Point score	Average	Average level
A	2B	3	2	51	17	2
B	4	4	4	81	27	4
C	5	3	4	81	27	4
D	2B	2	3	51	17	2
E	5	5	5	99	33	5
F	4	3	4	75	25	4
G	4	4	5	87	29	4
H	5	4	4	87	29	4
I	4	4	4	81	27	4
				Total 693		

Total points for the group = 693

Average for group = 693 ÷ 3 ÷ 9 = 25.7 (to 1 d.p.)

This equates to Level 4, as it lies in the range 24–30.

Spread

The data for Task 49 involved only a few pupils but they illustrate one of the problems with arithmetic means: stating just a single summary value involves a loss of information. This 'typical' value reveals nothing about the variety of different results that contributed to it. A group of pupils where everyone was Level 4 would also have an average Level 4, but would be very different from the group above, which contained one pupil at Level 2 and one at Level 5.

It would be more informative to also give some indication of the spread: average Level 4 with a spread from Levels 2 to Level 5.

Stating the arithmetic mean and range of a set of data provides only a limited view. For a large data set the range gives no indication of exactly how the values are spread across the range and more refined methods are needed.

> **Task 50   Look again**
>
> Look again at the point scores of the nine pupils considered in Task 49:
>
> 51   81   81   51   99   75   87   87   81
>
> Sort these into numerical order and find the median (mid score), the minimum value and the maximum value. Does this give you a better feel for the spread of results?

Comment
The median is 81.

51	51	75	81	81	81	87	87	99
↑				↑				↑
Minimum				Median				Maximum

This gives a feel for the spread by showing the lowest value, the middle value and the highest value. The most basic measure of spread is the range, calculated by calculating maximum – minimum. In this example:

range = maximum – minimum = 99 – 51 = 48.

One problem with the range is that it can easily be affected by an untypical value. For example, if one pupil had a point score of 21, the range would increase from 48 to 78 because of this single untypical result. One way around this is to measure the range between two values that are not quite at the extremes. The **interquartile range** measures the range of the middle half of the values. It is the difference between the values of the upper quartile and the lower quartile. These two **quartiles** are the values of the numbers that are one-quarter of the way in from either end of the batch once it has been sorted. It is possible to be more precise than this in defining a quartile but textbooks vary slightly as to how it is defined. It is enough here to know that, in order to find the quartiles you must first sort the data and then find the two values that are approximately one-quarter of the way in from each end.

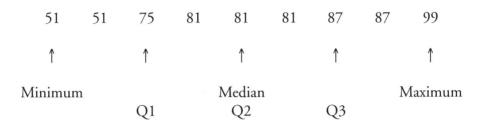

Figure 7 Quartiles illustrated using the data batch in Task 50

With the nine values in Task 50, the **lower quartile** is, roughly speaking, the value of the third number, while the **upper quartile** is, approximately, the value of the seventh number. The quartiles are often abbreviated to Q1 (the lower quartile) and Q3 (the upper quartile); the second quartile, Q2, is another name for the median.

A data batch can then be described by giving the following five-figure summary:

Minimum value	Lower quartile	Median	Upper quartile	Maximum value
Min	Q1	Med	Q3	Max

Boxplots

Boxplots, also known as box and whisker diagrams, are an effective way of displaying five-figure summaries. Figure 8 is a boxplot showing the point scores of the nine pupils in Task 50.

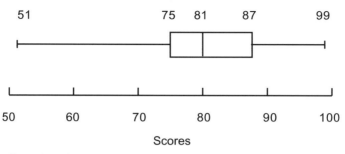

Figure 8 Boxplot showing the point scores of the pupils in Task 50

The 'box' that forms the central part of the boxplot represents the interquartile range; the middle half of the batch. The vertical line in the box represents the median.

This visual representation is particularly powerful for comparing several sets of data. The boxplot diagram can be presented horizontally or vertically.

Task 51 Using boxplots

A Key Stage 2 teacher was concerned about the reading progress of the boys in her class relative to the girls. She measured the reading ages of all the children in years and months, converted the data to decimal years and then found summaries of the data separately for girls and boys. These summaries are shown in Table 13.

Table 13 Reading ages

Summary value	Minimum	Lower Quartile (Q1)	Median	Upper Quartile (Q1)	Maximum
Girls	6.2	7.1	7.4	8.1	9.5
Boys	5.5	6.5	7.0	7.6	9.8

Display the two sets of results as horizontal boxplots, one below the other. Use the display to help decide whether she has good reason to be concerned about the boys' progress.

Comment

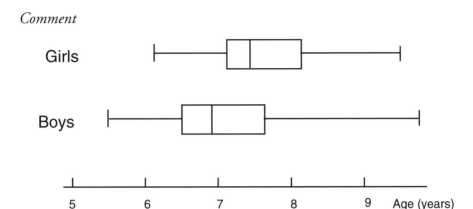

Figure 9 Boxplots showing girls' and boys' reading ages

Overall the girls' reading ages are higher, with one exception – the maximum value for the boys exceeds the maximum value for the girls. However, this could be explained by having just one exceptionally advanced boy in the class. At every other level of reading achievement, the girls outperform the boys by roughly half a year (i.e. by about 6 months). This suggests that the teacher does have reason to be concerned about the underperformance of the boys. Drawing the two boxplots together in this way allows her to see at a glance overall patterns in the data. For example, the lower quartile reading age for the girls (7.1 years) is actually higher than the median score for the boys (7.0 years).

Task 52 GCSE pass rates

The table shows the distribution of all 3137 mainstream schools in England as measured by the percentage of pupils gaining five or more GCSE Grades A*–E.

	5 %	Lower quartile	Median	Upper quartile	95 %
Percentage of pupils	15	30	44	58	91

Data taken from the 1999 Autumn Package

> To interpret these data it is easiest to think of all the 3137 schools being ranked in order according to the percentage of pupils gaining five or more GCSE Grades A*–E, the school with the lowest percentage on the left and the one with the highest on the right. The table shows that the lower quartile is 30%. This means that the school one-quarter of the way up has 30% of its pupils gaining five or more GCSE Grades A*–E (and that all schools below that have a lower pass rate).
>
> 1. What percentage of pupils gaining five or more GCSE Grades A*–E does the middle-ranked school have?
>
> 2. The upper quartile is 58%. Express in your own words the meaning of this.
>
> 3. The table also shows two other summaries. The 5% column gives the percentage pass rate for the school ranked 5% from the bottom. So the 5% of schools below that had less than 15% of their pupils gaining five or more GCSE Grades A*–E. Express in your own words the meaning of the 95% column.

Comment
1. 44%.

2. The school one-quarter of the way from the top had 58% of its pupils gaining five or more GCSE Grades A*–E. All schools above that had at least 58% pass rates.

3. The school ranked 95% from the bottom (i.e. 5% from the top) had a pass rate of 91%, so 91% of its pupils gained five or more GCSE Grades A*–E. All the schools in the top 5% had at least 91% of their pupils gaining five or more GCSE Grades A*–E.

As Task 52 showed, official data on national pupil performance are often given in more detail than just the five summary statistics. More values, called **percentiles**, are used. For example, the 30th percentile score is the score below which 30% of all the scores lie, and the 95th percentile score is that score below which 95% of all the scores lie. So, the lower quartile is the 25th percentile, the upper quartile is the 75th percentile and the 50th percentile is another term for the median (50% of the values lie below and above it).

National data comparing school performance, categorise schools by phase (primary, secondary unselected entry and secondary selected entry). Also, because social influences on the schools' intakes need to be taken into account, further indicators are used. The social indicator most commonly seen is the number of pupils eligible for free school meals. This may seem fairly crude, but it has been established as a reasonable measure of social deprivation. These data are published by the DfEE Standards Unit in the annual Autumn Package (see Resources section).

The intention of this data – the 'pupil performance benchmarking data' – is to allow schools to compare the spread of their pupil performance at each Key Stage with national results from similar schools. These measures are used as one basis for judging the quality of education provided. An individual school's performance is extracted from the national data to produce the annual PANDA (Performance AND Assessment) report and the PICSI (Pre-Inspection Context and School Indicator) report. Within LEAs, benchmarking data are also provided on how budgets are allocated within comparable schools.

Task 53 More boxplots

Figure 10 illustrates more data taken from the 1999 Autumn Package

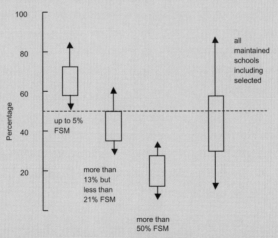

Figure 10 Boxplots showing the distribution of schools as measured by the percentage of pupils gaining five or more GCSE passes A*–C, where FSM means free school meals

(Note that the whiskers are shown with arrowheads, because the DfEE does not publish the extreme 5% of values.)

Use the boxplots in Figure 10 to answer the following questions.

1. Of the schools with up to 5% of pupils receiving free school meals (i.e. the left-hand boxplot), roughly what is the lower quartile percentage value? State in your own words what this means.

2. Of the schools with more than 50% of pupils receiving free school meals, roughly what is the upper quartile percentage value? State in your own words what this means.

3. Overall, what do these boxplots suggest to you?

Comment
1. The lowest quartile percentage value is about 59%. This means that, when considering this particular batch of schools, the school one-quarter from the bottom had 59% of its pupils gain five or more GCSE passes A*–C. Putting this another way, the three-quarters of these schools that are ranked above that school had 59% or more of their pupils gaining at least five or more GCSE passes A*–C.

2. The upper quartile percentage value is about 27%. This means that, with this particular batch of schools, the school one-quarter from the top had 27% of its pupils gaining five or more GCSE passes A*–C. All schools above this had at least a 27% pass rate. Putting this another way, roughly three-quarters of these schools had fewer than 27% of their pupils gain at least five or more GCSE passes A*–C.

3. There is a huge disparity in performance between schools with up to 5% free school meals and those with more than 50% free school meals. This suggests that social factors in the form of socioeconomic well-being are a strong indicator of examination success. What these boxplots do not do is reveal the exact mechanism that links poverty with examination performance.

Having such national information helps individual schools to investigate the reasons for their performance in relation to comparable schools. Relevant data are likely to include:

- pupil performance in different subjects;
- the relative performance of different groups of pupils, for example, boys and girls;

- those for whom English is not their first language;
- pupils who have been in the school for less than two years.

Once patterns can be identified, the school can target efforts to build on strengths and attempt to rectify weaknesses. Increasingly schools are being expected to undertake such detailed self-evaluation on a regular basis as part of whole-school performance management.

The DfEE actually publishes a seven-figure summary of performance figures. These include two extra percentiles, the 40th and the 60th percentiles. Rather than giving the maximum and minimum, their data show the 5th and 95th percentiles. The DfEE equates their figures to the following judgement of performance:

5%	25%	40%	50%	60%	75%	95%
	Lower quartile		Median		Upper quartile	
Well below national average		Just below	Average	Just above	Well above national average	

Other statistical summaries

Other statistical summaries are used to measure spread, such as mean deviation, variance and standard deviation, but these are not currently used by government agencies when publishing school and pupil performance data.

Representing and interpreting data in tables and graphs

Tables and graphs are used to present data in accessible and vivid forms. They enable users to spot trends, and see relationships and patterns much more readily than with lists of data. This section looks at various common types of tables and graphs.

Tables

A great deal of information is made available in tabular form; for example timetables, attendance registers, test scores, national summary data, benchmark indicators and league tables. Information is displayed in rows and columns, which if imported to a spreadsheet enables further summarising or processing (such as sub-totals, percentages) to be prepared easily.

Here is one of the more simple tables produced by the DfEE in its Autumn Package (Table 14). Interpreting such a table can be tricky. Task 54 gives you some practice.

Table 14 Recent trends in Key Stage 3 National Summary Results over the last five years for all pupils attaining Level 5 and above in English, mathematics and science statutory tests and teacher assessments

	Test					Teacher assessment				
	1996	1997	1998	1999	2000	1996	1997	1998	1999	2000
English	57	57	65	64	63	61	61	62	64	64
Mathematics	57	60	59	62	65	62	63	63	64	66
Science	57	60	56	55	59	60	62	62	60	62

Source: DfEE Autumn Package 2000.

Task 54 Reading a table

Use Table 14 to answer the following:

1. What do the data in the main body of the table show?

2. Which aspects show an increasing trend over the last five years?

3. Which aspect shows the greatest dip over the last five years?

Comment
1. They show the percentage of pupils attaining Level 5 and above in Key Stage 3 tests and teacher assessments over a five year period in the three subjects. It can be confusing to decide which is 'the body' of the table. This table has two main types of columns, each subdivided into five years, and three rows in the body. Notice that the information given does not actually say that the data in the body are percentages.

2. None of the aspects shows a consistently rising trend, although the teacher assessment results in English and mathematics do not show a fall.

3. Science test results show the greatest dip, of 4% between 1997 and 1998.

When reading tables try to focus on what information is included, and what is not included. The title as well as row and column headings provide essential information and should be read carefully.

Task 55 What does it mean?

Study Table 15 below and then try to answer the question: do boys get better results than girls in art at Key Stage 3?

Table 15 Percentage of boys and girls in England attaining each level in art teacher assessments in 2000

	Disapplied	Absent	A	B	C	D
Boys	0	1	29	52	15	3
Girls	0	1	15	51	27	6
All	0	1	22	52	21	4

Source: Based on data from Autumn Package 2000.

A represents pupils who are working towards the expectation for the end of Key Stage 3.
B represents pupils who are achieving the expectation for the end of Key Stage 3.
C represents pupils who are working beyond the expectation for Key Stage 3.
D represents pupils who are demonstrating exceptional performance.

Comment
In Categories A and B the percentage of boys is higher than that of the girls. But what is not clear just from the table is whether these are high or low grades. The notes give the information. In fact A is the lowest grade.

The categories are Key Stage descriptors. So at Key Stage 3 boys do not perform as well as girls in art.

Percentages can help to make differences easier to see. However, as Task 56 shows, they can be used inappropriately and give a misleading impression.

Task 56   Absent

Look at Table 16. It provides a summary of unauthorised absences tabulated by year group and term, along with term and year group totals.

Table 16   Unauthorised absences

Year group	No. of pupils	Term 1	Term 2	Term 3	Total	Percentage of unauthorised absences
7	238	36	41	42	119	$\frac{119}{388} = 0.31 = 31\%$
8	145	22	36	48	106	
9	136	24	32	28	84	
10	138	25	12	18	55	
11	143	4	12	8	24	
Total	800	111	133	144	388	100

Percentages are rounded to the nearest whole number.

The final column is found by calculating the total absences for each year group as a percentage of the total number of unauthorised absences for the school. Using the first calculation as a guide, find the percentage of the total number of unauthorised absences for each year group. Where does the school have a possible problem?

Comment

Table 17 Completed unauthorised absences table

Year group	No. of pupils	Term 1	Term 2	Term 3	Total	Percentage of unauthorised absences
7	238	36	41	42	119	$\frac{119}{388}$ = 31%
8	145	22	36	48	106	$\frac{106}{388}$ = 27%
9	136	24	32	28	84	$\frac{84}{388}$ = 22%
10	138	25	12	18	55	$\frac{55}{388}$ = 14%
11	143	4	12	8	24	$\frac{24}{388}$ = 6%
Total	800	111	133	144	388	100

These calculations suggest that the biggest problem that the school has with unauthorised absences is with the Year 7 class (at 31%). However, this does not take into account the fact that, the number of Year 7 pupils (238) is much larger than any other year group. It should therefore not be surprising that the actual number of unauthorised absences in Year 7 is larger for this reason alone. A more useful calculation is to find the unauthorised absence rate for each year group, based on the number of half-days that each year group could register. This is calculated for Year 7 as follows.

There are 380 pupil half-days in a school year so for a year group of 238 pupils the maximum number of possible registrations is 238 × 380 = 90 440. Since the actual number of absences for Year 7 was 119, the Year 7 unauthorised absence rate for that year was: $\frac{119}{90\,440}$ = 0.13 = 13%.

Task 57 Recalculating the percentages

Table 18 shows unauthorised absence rate. Using the first calculation as a guide, find the unauthorised absence rate for each year group, giving percentages to 2 d.p. Where now do you feel that the school has a possible problem?

Table 18 Unauthorised absences

Year group	No. of pupils	Term 1	Term 2	Term 3	Total	Percentage of unauthorised absences
7	238	36	41	42	119	$\frac{119}{238 \times 380} = 0.13\%$
8	145	22	36	48	106	
9	136	24	32	28	84	
10	138	25	12	18	55	
11	143	4	12	8	24	
Total	800	111	133	144	388	

Comment

The recalculated percentages tell a different story. The Year 7 pupils do not pose as large a problem in this area as do the pupils in Years 8 and 9 (Table 19).

Table 19 Completed unauthorised absences table

Year group	No. of pupils	Term 1	Term 2	Term 3	Total	Percentage of unauthorised absences
7	238	36	41	42	119	$\frac{119}{238 \times 380} = 0.13\%$
8	145	22	36	48	106	$\frac{106}{145 \times 380} = 0.19\%$
9	136	24	32	28	84	$\frac{84}{136 \times 380} = 0.16\%$
10	138	25	12	18	55	$\frac{55}{138 \times 380} = 0.10\%$
11	143	4	12	8	24	$\frac{24}{143 \times 380} = 0.043\%$
Total	800	111	133	144	388	

The rates in Years 8 and 9 do raise questions. Are there a few disaffected pupils, or are there a number of pupils who are careless about getting parental notes that would enable the school to authorise their absences? These questions could only be answered by going back to the primary data: the attendance registers.

Task 58 Absence rates

The national secondary school average of unauthorised absence was 1.1% in 1999. Calculate the unauthorised absence rate for the school in Tasks 56 and 57.

Does the school in Tasks 56 and 57 have an unauthorised absence problem overall?

Comment

There are 380 pupil half-days in a school year, so for a school of 800 pupils there were 800 × 380 = 304 000 pupil half-days, of which 388 were unauthorised absences (from Table 19).

The rate is therefore 388 ÷ 304 000 × 100% = 0.13%.

This is considerably lower than the national average, and so is not a major problem.

Remember, when 'reading' tables try to focus on what information is included, and what is not included. The title as well as row and column headings provide essential information and should be read carefully.

Task 59 Progress

In March of a particular year, a group of pupils were given a standardised reading test. Those with a reading age at least 6 months lower than their actual age were given extra reading support. The group was retested in July.

Table 20 shows the scores for five of the pupils. Use the information to decide whether the extra support strategy was worthwhile.

Table 20 Reading test results Year 1 group

Pupil	1 March			1 July			Gain in months
	Actual age	Reading age	Difference in months	Actual age	Reading age	Difference in months	
A	5-11	5-5			6-0		
B	6-1	7-4			7-7		
C	5-8	6-0			6-2		
D	6-6	5-11			6-5		
E	5-10	5-11			6-0		

Note: Reading ages are given in years-months, or sometimes confusingly written years.months.

Comment

Table 21 Completed reading test results Year 1 group table

Pupil	1 March			1 July			Gain in months
	Actual age	Reading age	Difference in months	Actual age	Reading age	Difference in months	
A	5-11	5-5	−6	6-3	6-0	−3	+3
B	6-1	7-4	+15	6-5	7-7	+14	−1
C	5-8	6-0	+4	6-0	6-2	+2	−2
D	6-6	5-11	−7	6-10	6-5	−5	+2
E	5-10	5-11	+1	6-2	6-0	−2	−3

Only pupils A and D would had been given extra reading support (Table 21). Pupil A has improved by 3 months and pupil D by 2, so the strategy would seem to have been successful for these pupils. However, the other pupils have not progressed as much as might be expected in 4 months, which raises questions about how the strategy was implemented.

Frequencies

The detail of raw data can be of interest, but often the amount of information can obscure any overview. For example, a class achieved the following percentage marks in an examination:

Girls 23 55 71 42 66 11 45 77 32 61 37 55 51 87 61

Boys 68 49 95 68 43 56 77 43 55 89 66 51 60 56 51

This collection of unordered, raw data gives no immediate insight into the performance of individuals in relation to the whole class. One useful organising device is known as a **stemplot** (also known as a **stem-and-leaf** diagram). In this case the 'stem' is the 'tens' digit and the 'leaves' are the 'units' digit of each mark. For example, the mark of 23 by one of the girls is represented with the units digit (the '3') placed on the row corresponding to 2 tens (Figure 11).

```
1:  1
2:  3
3:  2 7
4:  2 3 3 5 9
5:  1 1 1 5 5 5 6 6
6:  0 1 1 6 6 8 8
7:  1 7 7
8:  7 9
9:  5
```

Figure 11   Stemplot

The visual aspect of this type of table makes it almost a diagram: you can easily see where the test scores are in relation to one another. Notice that the 'leaves' have been put in order to see the actual marks within each ten more readily.

Task 60   Stemplot

Even more of the information about the test scores can be retained by a 'back-to-back' stemplot. Some of the data from the class scores are shown in Figure 12 with the first three boys' values (68, 49 and 95) on the left and three of the girls' values (23, 55 and 71) on the right.

Complete the back-to-back stemplot for the girls' and boys' marks. What does it reveal about these two sets of marks?

```
        Boys      Girls
             :1:
             :2:  3
             :3:
           9 :4:
             :5:  5
           8 :6:
             :7:  1
             :8:
           5 :9:
```

Figure 12   Back-to-back stemplot

Comment

The final back-to-back stemplot is shown in Figure 13. Again, the leaves on each level have been sorted into ascending order, making it easier to identify particular values such as the girl with the third highest score or the median boys' score.

```
            Boys        Girls
              :1:  1
              :2:  3
              :3:  2 7
       9 3 3  :4:  2 5
     6 6 5 1 1 :5:  1 5 5
       8 8 6 0 :6:  1 1 6
              7 :7:  1 7
              9 :8:  7
              5 :9:
```

Figure 13 Completed back-to-back stemplot

Overall, the average scores for boys and girls are fairly similar but the diagram shows clearly that the spread of girls' marks is wider than the spread of boys' marks and they contain more lower scores.

It is a short step from a stemplot to produce a frequency table that groups the marks achieved. The term 'frequency' refers to the total number of children within a group mark range.

Task 61 Frequency table

Complete the frequency table (Table 22) of the test scores from Task 60. You may find it easiest to use the data from the stemplot.

Table 22 Frequency table

Mark range	Frequency		
	Girls	Boys	All
0–9	0	0	0
10–19	1	0	1
20–29			
30–39			
40–49			
50–59			
60–69			
70-79			
80-89			
90-100			

Comment

Table 23 Completed frequency table

Mark range	Frequency		
	Girls	Boys	All
0–9	0	0	0
10–19	1	0	1
20–29	1	0	1
30–39	2	0	2
40–49	2	3	5
50–59	3	5	8
60–69	3	4	7
70–79	2	1	3
80–89	1	1	2
90–100	0	1	1

Sometimes the actual frequencies do not give a sufficiently illuminating picture, and it is necessary to process them further. For example, consider the much larger set of test scores shown in Table 24.

Table 24 Test score frequencies

Score	Frequency
0–9	0
10–19	5
20–29	10
30–39	15
40–49	30
50–59	35
60–69	25
70–79	20
80–89	15
90–100	5

Here there are a total of 160 pupils (found by adding up all of the frequencies). If we were to ask 'What proportion had scores in the 40s?' it is not easy to answer, as most of us cannot easily recognise proportions of 160. In such cases converting the frequencies to percentages makes the data easier to interpret. Task 62 asks you to do this.

Task 62   Percentage frequencies

For the data in Table 25, since the total frequency is 160, the percentage frequency is found by dividing the frequency by 160 and converting to a percentage corrected to a whole number.
Complete Table 25.

Table 25 Frequency and percentage frequency table

Score	Frequency	Percentage frequency
0–9	0	0%
10–19	5	3%
20–29	10	
30–39	15	
40–49	30	
50–59	35	
60–69	25	
70–79	20	
80–89	15	
90–99	5	

What percentage of pupils scored 40–49?

Comment

Table 26 Completed frequency and percentage frequency table

Score	Frequency	Percentage frequency
0–9	0	0%
10–19	5	3%
20–29	10	6%
30–39	15	9%
40–49	30	19%
50–59	35	22%
60–69	25	16%
70–79	20	13%
80–89	15	9%
90–99	5	3%

So 19% of pupils scored 40–49.

Cumulative frequency

When you want to answer questions such as 'How many, or what percentage of pupils achieved more than or less than a certain value?' it is useful to have data in cumulative form. Much of the data produced nationally is presented in this way. Consider the frequency table of class scores used in Task 60 (Table 27).

Table 27 Frequency table of class scores from Task 60

Mark range	Frequency	Cumulative frequency
0–9	0	0
10–19	1	1
20–29	1	2
30–39	2	4
40–49	5	9
50–59	8	17
60–69	7	24
70–79	3	27
80–89	2	29
90–100	1	30

The new column, cumulative frequency, shows the total number of pupils scoring up to that point. For example, in the row

40–49	5	9

the cumulative frequency, 9, shows that nine pupils scored 49 or less. The last line shows, unsurprisingly, that 30 pupils (the whole class) scored 100 or less.

As with frequencies, it is often more useful to have the cumulative frequencies given as percentages. Task 63 asks you to work with percentage frequencies and cumulative frequencies.

Task 63   Cumulative frequencies

Table 28 shows the test data for the large group given in Table 26, with the cumulative column now showing the cumulative percentage frequencies to the nearest whole number.

Table 28   Test data from Table 26, and cumulative percentage frequencies

Score	Frequency	Percentage frequency	Cumulative percentage frequency
0–9	0	0%	0%
10–19	5	3%	3%
20–29	10	6%	9%
30–39	15	9%	18%
40–49	30	19%	37%
50–59	35	22%	59%
60–69	25	16%	75%
70–79	20	13%	88%
80–89	15	9%	97%
90–99	5	3%	100%

What percentage of pupils scored:

1. 40–49
2. 49 or less
3. 80 or more?

Comment
1. 19% of pupils scored 40–49 (read off from the frequency column).
2. 37% of pupils scored less than 50 (read off from cumulative frequency column).
3. 12% of pupils scored 80 or more (88% scored less than 80, so 100% − 88% scored 80 or more).

You have met the percentage cumulative frequencies before: they are the percentiles, including the quartiles and median (see p. 80). In Table 28 the cumulative percentage frequency up to a score of 69 is 75%. That is the upper quartile: one-quarter of the pupils from the top.

When the percentage cumulative frequencies are plotted on a graph, the **cumulative frequency graph** (Figure 14), it is easy to find the median, quartiles and other percentiles.

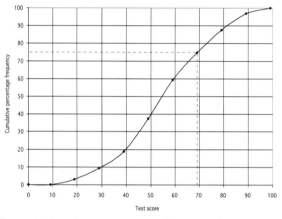

Figure 14 Cumulative frequency graph, sometimes called an **ogive**

Note that the points on a cumulative frequency graph are plotted at the end of each interval. For example, the 75% value is plotted at a score of 69.

The percentiles, quartiles and the median scores can be read off from the graph. For example, to read off the median go across at 50% frequency until the graph line is reached and then drop vertically to read off the score (55).

Task 64   Reading off frequencies

Using the cumulative frequency graph in Figure 14 answer the following:

1. What is the lower quartile score?

2. What percentage of pupils scored less than 40?

Comment

The scale of the graph makes it difficult to read off precise percentages but it is possible to estimate.

1. Approximately 45 (the lower quartile is the 25% cumulative frequency).

2. Just under 20% (this involves reading off in the opposite direction – up from the score to the graph and then horizontally to read off percentage cumulative frequency.)

More on charts and graphs

It is relatively easy to read data when they are presented in a table, and when they are put in a spreadsheet they can be further processed to produce summary data. However, tables are not always the best way of presenting information, nor are they a form from which it is particularly easy to identify relationships or trends. Charts and graphs are often much more useful for gaining an overview of the data, but their drawback is that yet more detail is lost.

What is the distinction between a chart and a graph? Very broadly, a chart illustrates data whereas a graph can be used to read off further information.

Task 65 What is lost?

The information given in Task 56, given again in Table 29, summarises the unauthorised absences in a school, tabulated by year group and term, along with the term and year group totals.

Table 29 Unauthorised absences

Year group	No. of pupils	Term 1	Term 2	Term 3	Total
7	238	36	41	42	119
8	145	22	36	48	106
9	136	24	32	28	84
10	138	25	12	18	55
11	143	4	12	8	24
Total	800	111	133	144	388

This data can be presented as a bar chart (Figure 15).

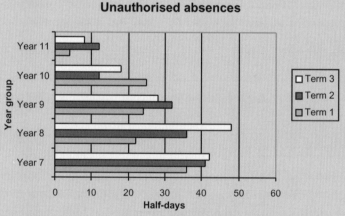

Figure 15 Bar chart showing unauthorised absences

This presentation highlights the Year 8 differences more clearly than the figures alone, but what information is lost if only the bar chart is made available?

Comment

As with putting data into a table, making a graphical representation stresses some aspects but makes other aspects less apparent (detail may be lost). In this case the number of pupils in each year group is not available from the chart; so the bigger Year 7 group appears to be as much of a problem as Year 8 (like is not being compared to like).

If only the chart was available, the opportunity to calculate the percentage of unauthorised absences to pupil half-days would also be lost. This may not seem important, but that percentage indicated that the overall unauthorised absences was not a significant problem, even in Year 8 (see p. 89.)

As with tables, graphs and charts need to be read with some care. Titles and axes labels should give a clear idea of what data are included and the units being used. In the media and sometimes even in official statistics the labels and titles are sometimes omitted. This means that some preparatory work needs to be done to ensure that the wrong conclusions are not drawn.

Bar charts

Bar charts are frequently used to present *discrete* data, that is, data that have been categorised and presented as a straight count or processed to percentages. The bars can be presented either horizontally or vertically, but the length of bar or height of columns is the 'count' (compared to a histogram, see p. 106, where the area represents the frequency). It is the relative heights or lengths that are being stressed. By convention there should be a gap between the bars of individual categories but this is not always adhered to in the media. Where the actual length of the bar may be significant or may be needed for further processing the relevant figures are placed on the chart (Figure 16).

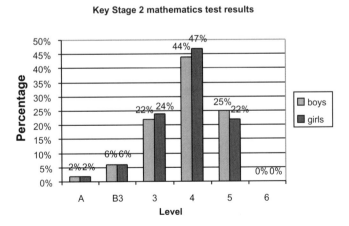

Figure 16 Bar chart showing Key Stage 2 mathematics test results

Task 66 On track?

Look again at Figure 16, which is based on data from the Autumn Package 1999.

In 1998 the government set the target that 75% of 11-year-olds would reach Level 4 by 2002.

What percentage reached Level 4 in 1999?

Is there anything you do not understand from the data as presented?

Comment

Boys reaching Level 4 and above = 44% + 25% = 69%

Girls reaching Level 4 and above = 47% + 22% = 69%

So 69% of both boys and girls reached Level 4 or above in 1999.

Categories A and B3 are not self-explanatory on the chart. A is the percentage of pupils absent; B3 is the combined figures for those awarded Level 2 and below.

Statistics produced for one purpose are frequently used for another. For example, national pupil performance data are used for a variety of purposes, such as comparing schools, providing benchmarking data and value added calculations.

By analysis of individual pupil performance in Key Stage SATs tests, the DfEE has produced data on the chances of pupils obtaining particular levels at the next stage, based on their achievement at the previous one. Thus data that were collected for purposes of comparison are being used for a different purpose, that of prediction.

Task 67 Chances

The data in the Autumn Package include pupil performance graphs, which compare pupil performance at one Key Stage with what they achieved in the previous one. They are produced from matched pupil records.

1. Explain in your own words what the bar chart in Figure 17 shows.

2. How might the information be used?

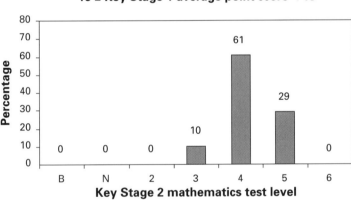

Figure 17 Bar chart

Comment
Looking at the title, this graph gives information about pupils whose average point score at Key Stage 1 was at least 16 and less than 18, i.e. a high Level 2. The horizontal axis indicates that it refers to Key Stage 2 mathematics test scores. The vertical axis is the percentage achieving the various levels. What the chart shows is that of the pupils who scored at least 16 and less than 18 at Key Stage 1, 10% achieved Level 3, 61% Level 4 and 29% Level 5 in the Key Stage 2 mathematics tests.

The information could be used to guide target setting for current pupils (hence 'chances'). For example Level 4 should be considered the 'norm', but Level 5 would be an 'achievable but challenging' target for at least some pupils. But it is also a benchmark of average national achievement against which a school can measure its own past performance.

Histograms

Histograms are used to represent frequencies of continuous data, such as pupils' ages, heights, etc. In a histogram it is not the height of a column that gives the frequency, but the *area* of each column.

Task 68   Histogram or bar chart?

Look at the histogram in Figure 18, which has been drawn from data of pupil times on a cross-country run.

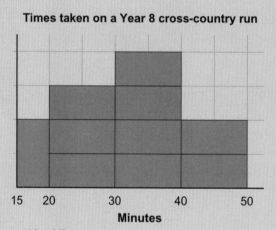

Figure 18   Histogram

1. What percentage of the pupils finished the run in less than 40 minutes?

2. What general features distinguish the histogram from a bar chart?

Comment
1. The total shaded area represents the whole population, i.e. 100% of the Year 8 pupils who took part. The 40–50 minute column is 20% of the area, so 80% finished in less than 40 minutes. Another way of thinking about this is that the 15–20 minute column represents 10% of the pupils, the 20–30 minute column is three times that area so represents 30%, and the 30–40 minute column is four times the area so represents 40%. Totalling these gives 80%.

2. There are several important differences between a bar chart and a histogram. Three of them are discussed here:

(a) *There are no gaps*: Because a histogram is used to represent continuous data, there is no gap between where one interval ends and the next one begins. For example, the column 30–40 minutes actually covers the range of times 30 minutes and 0 seconds up to 39 minutes and 59 seconds. If times are measured to the nearest tenth of a second, this interval would cover the range of times 30 minutes and 0.0 seconds up to 39 minutes and 59.9 seconds. Similarly, the adjacent column labelled 40–50 actually covers the range of times 40 minutes and 0 seconds up to 49 minutes and 59 seconds. There is therefore no gap between the columns: they touch at the value 40 minutes 0 seconds.

(b) *Columns are of unequal width*: Notice that the columns are not all of equal width (the first column represents a five minute interval whereas the others show ten minute intervals). Although histograms are often drawn with equal width columns, they can also be drawn with unequal columns as shown here. This is in contrast with the drawing of bar charts, for which equal bar widths is an essential requirement.

(c) *There is no vertical scale*: Notice that this histogram does not include a scale on the vertical axis. The reason is that such a scale would be ambiguous, especially in a case like this where the columns are of unequal width. The next task looks further at this issue.

Task 69 Why no scale on the histogram?

Figure 19 shows an *incorrectly drawn* version of the previous histogram with a 'Frequency' scale included on the vertical axis. Look at this histogram and try to answer the following questions, which aim to point up the ambiguity of such a scale.

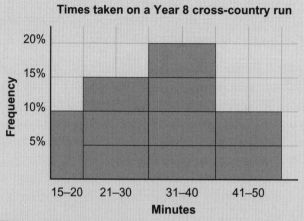

Figure 19   Incorrectly drawn histogram

1. What is the frequency of pupils in the 15–20 minute interval?

2. What is the frequency of pupils in the 41–50 minute interval?

3. In what way are your answers to 1 and 2 inconsistent?

Comment
1. and 2. Reading off the vertical scale gives the same answer: 10%.

Clearly there is an inconsistency here since the area corresponding to the 41–50 minute interval is twice as large as that of the 15–20 minute interval.

The main point to emerge here is that, when drawing histograms, it is the *area* of each column that indicates the frequency, and not the height. This means that providing a numerical scale on the vertical axis should be avoided, as there is scope for confusion as to what such a scale would mean. (Some presentations use a numerical scale but labelled 'frequency density'.)

The one exception to this is where the class intervals are all equal, in which case a vertical scale indicating the frequency can be included.

Pie charts

A pie chart is suitable for representing named data, i.e. categories of things, rather than numerical data. A pie chart is drawn so that the size of each slice of the 'pie' is a measure of the amount of that category.

Table 30 shows the numbers of school pupils by type of school in the public sector in 1998–99.

Table 30 School pupils in the public sector (source: *Social Trends* 30)

Type of school	Number of pupils in thousands
Nursery	109
Primary	5376
Secondary modern	92
Grammar	203
Comprehensive	3205
Other	291
All public sector schools	9276

Using a computer it is easy to represent these figures as a pie chart. However, without intelligent input from you, the computer will not always produce a sensible pie chart, as you will see with Task 70.

Task 70 A silly pie chart
The pie chart in Figure 20 has been drawn from the data in Table 30. However, the chart is not sensible. Can you see why?

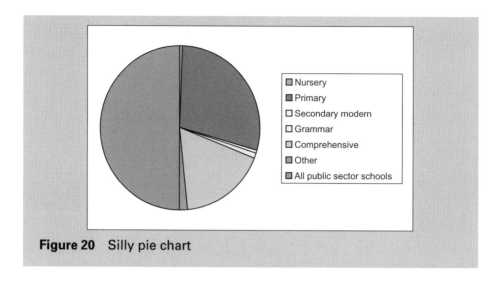

Figure 20 Silly pie chart

Comment

There are two problems here, one major and one minor. The big mistake is the inclusion of the slice labelled 'All public sector schools'. As this represents the sum of all the other categories, it makes no sense at all to include them again lumped into a single huge category.

A second, less serious problem is the inclusion of some very small categories such as 'Nursery' and 'Secondary modern'. This produces slices of the pie that are really too small to be clearly visible and so they might usefully be included with one of the other categories. A redrawn pie chart taking account of these two issues is shown in Figure 21, based on the adapted data in Table 31.

Table 31 School pupils in the public sector

Type of school	Number of pupils in thousands
Primary and nursery	5485
Grammar and secondary modern	295
Comprehensive	3205
Other	291

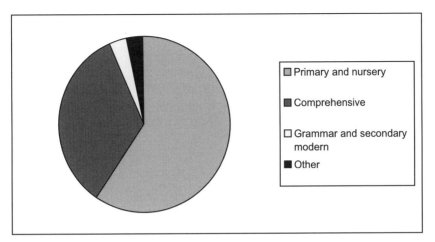

Figure 21 Sensible pie chart

One final point worth making about a pie chart is that it usually depicts data for which the order in which the categories are placed is not of any significance. Indeed, if the order is important then a bar chart would be a better choice, since the horizontal axis would preserve the impression of the ordering better than the circular arrangement of a pie chart. Nevertheless, there is one useful convention about the ordering of the slices of a pie chart, namely that they be presented in decreasing order of size, starting clockwise from '12 o'clock' on the pie. Look carefully at the above and you will see that the sectors have been re-ordered in this way.

Task 71 provides another example of a pie chart being used inappropriately.

Task 71 Why a pie chart?

Look at Table 32 and the pie chart in Figure 22 and try to decide why a pie chart is inappropriate. What would have been a better representation for this particular set of data?

Table 32 Classes of more than 30 pupils, 1998–99

Country	Percentage
England	18
Wales	11
Scotland	16

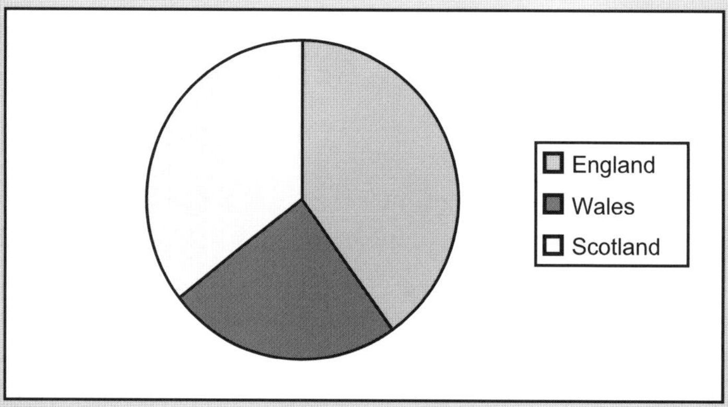

Figure 22 Inappropriate pie chart

Comment
Again, this is a rather silly use of a pie chart. The problem is that *the three slices taken together* representing the complete pie *are meaningless*. A key feature of a pie chart then is to stress the proportions of a whole. These data would be more sensibly represented as a bar chart, as in Figure 23.

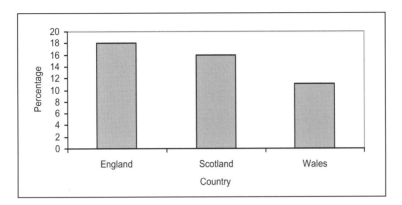

Figure 23 Percentage of classes with more than 30 pupils 1998–99

Note that the three countries have been re-ordered by size from largest to smallest. This is another useful convention that helps the reader to see overall patterns more clearly.

Emphasis on the comparative areas of the various sectors of a pie chart is the main point of Task 72.

Task 72 Report to parents

A small primary school wants to explain to parents how the non-teaching items of the school budget are spent. A pie chart does this well, so they provided parents with the representation in Figure 24.

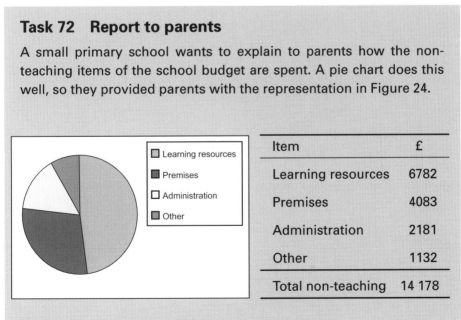

Item	£
Learning resources	6782
Premises	4083
Administration	2181
Other	1132
Total non-teaching	14 178

Figure 24 Non-teaching items of the school budget

1. If your aim as a teacher, parent or governor is to get a rough idea of the relative proportions spent on each category, what overall patterns do you see?

2. Which of the two representations above – table or pie chart – make it easier to draw these sorts of conclusions?

Comment

Learning resources use up just under a half of the non-teaching expenditure, while premises take just over a quarter. Most people would find it easier to see these overall patterns from the pie chart than from the table. However, if you need to know the precise figures, the table provides essential information.

The best of both worlds is to use the facility of including data on the chart itself (Figure 25).

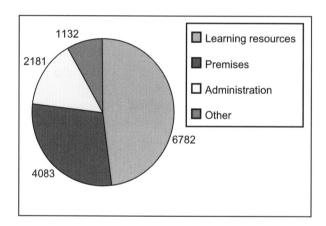

Figure 25 Non-teaching items of the school budget

Line graphs

Line graphs are often used for showing changes in data over time. By convention, time is always represented on the horizontal axis. For example, the graph in Figure 26 shows room temperature over a day. As you can see, the temperatures were monitored hourly, on the half-hour. The points are joined by a dotted line, indicating that any intermediate values read off here are estimates only.

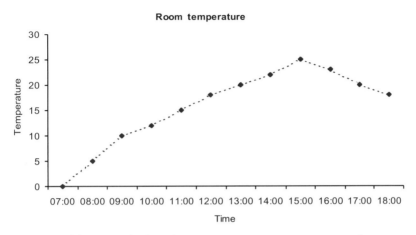

Figure 26 Line graph showing room temperature over a day

Note, the lines on the graph are dotted to indicate a trend; strictly, intermediate temperatures cannot be read off e.g. the temperature at 12 noon.

Task 73 Reading temperatures

1. Describe the overall pattern in classroom temperature during the course of the day.

2. Do you think that these data represent the room temperature of a real classroom or are they made up?

Comment
1. The graph indicates a steady temperature rise until 3.30pm and then a gradual falling off.

2. The data are invented (although those who have worked in old temporary buildings may not have thought so). You can tell from the fact that the classroom is so cold at the start of the day. Most teachers wouldn't work in temperatures of 5°C at 8.30am! Also, with the advent of thermostats and blinds, classroom temperatures are likely to remain fairly even throughout the teaching day, while these figures show a range of roughly 20°C.

Scatter graphs (scatterplots)

Data sometimes come in pairs; that is, for each item or person being considered *two* measurements are made rather than just one. For example, a group of pupils might undergo a health check where their heights *and* weights are measured. Alternatively they may undergo a class test where their reading age *and* their performance on a verbal reasoning test are assessed. When paired data are displayed graphically this two-dimensional aspect needs to be depicted and a scatterplot (often called a scatter graph or scatter diagram) does this very well.

Table 33 shows the marks in mathematics and English for eight students.

Table 33 Marks in mathematics and English

Student	Mathematics	English
Alison	53	66
Beena	62	53
Curtis	45	44
Denny	55	61
Eva	79	68
Fatima	35	40
Gary	62	65
Heather	72	70

The scatterplot in Figure 27 shows these data. Notice that each point on the graph corresponds to a particular *person*.

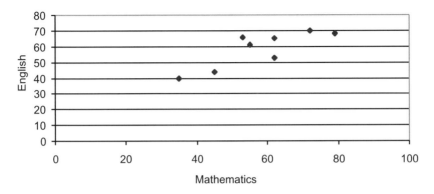

Figure 27 Scatterplot

Task 74 Following the plot

1. Spend a few moments identifying which point corresponds to which person. For example, the point that is furthest to the right corresponds to Eva, with a mark of 79 in mathematics and 68 in English.

2. What would it mean to join up the points on the scatterplot?

3. How would you describe the overall relationship between mathematics marks and English marks for these students?

4. An additional student, Ira, had a mark of 63 in mathematics. How could you use this scatterplot to predict Ira's likely mark in English?

Comment

1. Reading from left to right, the points belong to Fatima, Curtis, Alison, Denny, Gary (higher point), Beena (lower point), Heather and Eva.

2. Joining up the points is a meaningless exercise since there is no natural order in the points; remember that they simply represent a collection of people! Contrast this with the line graph where there was a progression over time from one point to the next. What is more useful is to see the overall trend in the relationship between the two

variables in question and this is achieved by drawing in the 'line of best fit' (Figure 28).

3. There seems to be a fairly close *positive* relationship between mathematics marks and English marks. In other words, students who did well in mathematics also did well in English, and vice versa. If students who did poorly in English did well in mathematics (or vice versa) this would indicate a *negative relationship*.

4. Using the scatterplot as a tool for prediction is a common procedure but care must be taken not to make excessive claims about the accuracy of the prediction. The usual procedure is to draw a line of best fit through the scatter of points to indicate the overall trend. This is shown in Figure 28. Another pupil, Ira's mathematics mark, 63, can then be read off the 'best fit' line and a prediction made of his English mark (roughly 65).

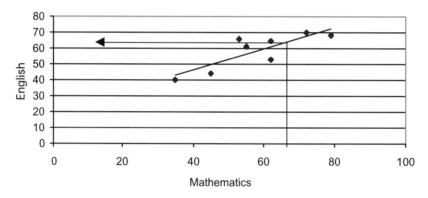

Figure 28 Scatterplot and line of best fit

A common mistake when interpreting scatterplots is to believe that a close relationship between two things (often referred to as a close *correlation*) proves that one has caused the other. Proving cause and effect is actually very difficult, as Task 75 shows.

Task 75   Cause and effect

A survey of 20 children aged between 5 and 15 revealed very close association between the amount of weekly pocket money they received and their height. How do you interpret this finding?

Comment
It might be tempting to deduce that children will therefore grow more quickly if you increase their pocket money, but this is clearly nonsense! There is a third factor here, which is the *age* of the children. Older children tend to receive more pocket money and older children tend to be taller. To claim a direct cause-and-effect relationship between pocket money and height is silly! Nevertheless, it is surprising how often this fallacy is trotted out in education. For example, many people believe that more testing results in better pupil performance, whereas there may be many other better reasons for improvements in measured performance (pupils are given more practice at tests, teaching has become more focused, and so on). There is some truth in the old adage that measuring the size of the apples in the orchard every day doesn't necessarily make for bigger or more flavoursome apples!

Using performance indicators and benchmarking data

In the Autumn Package, the Standards and Effectiveness Unit (SEU) of the DfEE present information that they claim enables schools to calculate their '**value-added**', meaning the progress made between two Key Stages. This information consists of national data on what pupils with particular scores at any one Key Stage have achieved at their next one. The data are given as bar charts (see Figure 17 p. 105) and as line graphs. Producing the data as bar charts requires a lot of charts (one for each range of average points scored) whereas all the information can be presented on one line graph, although some detailed information is lost.

Figure 29 shows the line graph for pupils progressing from Key Stage 3 to Key Stage 4.

The graph is produced by relating the average Key Stage 3 SATs results points (English, mathematics and science) to the average points achieved for GCSE/GNVQ results of the *same* pupils for all subjects (i.e. matched pupil data). The solid line is the median and the dotted lines the upper and lower quartiles.

Graph 3.1 1999 GCSE / GNVQ Total Point Score Median Line (with Quartile Boundaries)

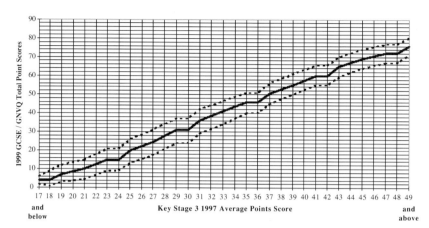

Figure 29 Line graph showing pupils progressing from Key Stage 3 to Key Stage 4, as presented by the DfEE SEU (source: Autumn Package 1999)

Looking at the pupils who score 36 points at Key Stage 3, then the graph shows that their median points score at GCSE/GNVQ is 46, with the lower quartile 40 and upper quartile 51. This shows that, of those pupils who scored 36 points at Key Stage 3, the middle half of them scored between 40 and 51 points at GCSE/GNVQ.

But what does this mean? To interpret this graph, you need to know what is meant by the 'points score'. How points are scored is shown in Tables 34–6.

Table 34

Key Stage 3 level	Points
3	21
4	27
5	33
6	39
7	45

Table 35

GCSE grade	Points
G	1
F	2
E	3
D	4
C	5
B	6
A	7
A*	8

Table 36

GNVQ		Points for		
		Pass	Merit	Distinction
Part 1	Foundation	3	6	8
	Intermediate	10	12	15
Full	Foundation	6	12	16
	Intermediate	20	24	30
Language unit	Foundation			2
	Intermediate			3.5

> **Task 76 Value-added**
>
> Use the data in Tables 34–6, and Figure 29 to answer the following questions.
>
> 1. If a pupil achieves an average of Level 5 at Key Stage 3, how likely is (s)he to get five GCSEs at grade C or higher?
>
> 2. How might the graph in Figure 29 be used to judge an actual school's 'value-added'?

Comment
1. Level 5 is 33–6 points, and five GCSEs at C or above is at least 30 points. From the graph, a bare 33 points at Key Stage 3 has, in the past, led to a lower quartile of 34 points, a median of 41, and an upper quartile of 47 at Key Stage 4. So, 75% of pupils who achieved a bare average Level 5 at Key Stage 3 did reach the equivalent average of five GCSEs at Key Stage 4. However, these are average results, the Key Stage 3 SATs are only for three subjects and the GCSE results are for all subjects. So a pupil with a bare average Level 5 is not certain to get five GCSEs, but it is possible.

2. By using a copy of the graph and superimposing the school's own pupil matched results it would be possible to identify whether the school performed better or worse than the overall national picture at Key Stage 4.

The Autumn Package contains a great deal of national school and pupil performance data for each Key Stage and for each subject that is assessed by national tests or examinations. To make school analysis easier, the Autumn Package is also available in interactive form. The software enables the input of individual school and pupil data and then produces graphs showing this in comparison to national data. It can also produce trend lines over several years.

> **Task 77   What use?**
>
> The national school and pupil data are used by OFSTED, LEAs and school management to make comparisons and judgements, but what use do you think they might be to an individual class teacher?

Comment
Understanding and using national pupil performance data may assist a class teacher in:

- setting realistic yet challenging targets for pupils;
- explaining predictions to pupils and parents;
- monitoring and evaluating his or her own and pupil performance;
- identifying particular groups of pupils who may be cause for concern.

Some LEAs are now producing local matched data in the form of lines of best fit rather than the median and quartile form of the national data. These are easier to understand and use to make comparisons with class or whole-school data.

> Increasingly, to be an effective teacher it is necessary to understand the possibilities *and* the limitations of numerical and graphical information. Every teacher now needs to be professionally numerate.

Resources

Websites

The precise addresses (URLs) of websites can be unreliable due to websites being updated, reorganised, or even disappearing altogether! If this happens, try using key words in a web search engine.

Test contexts: www.canteach.gov.uk/info/skillstests/numeracy/test coverage.htm (accessed March 2001)

Example and practice questions: www.canteach.gov.uk/info/skillstests/numeracy/questions (accessed March 2001)

www.canteach.gov.uk/info/skillstests/numeracy/questions/set2/

DfEE Autumn Packages on pupil performance: www.standards.dfee.gov.uk/performance/(accessed March 2001)

Glossaries of mathematical terms

QCA   www.qca.org.uk/ca/subjects/mathematics/maths_vocabulary.asp (accessed March 2001)

TTA: www.canteach.gov.uk/info/skillstests/numeracy/glossary/p.htm (accessed March 2001)

Books

Graham, Alan (1999) *Teach Yourself Statistics*. London: Hodder and Stoughton.
Graham, Alan (1995) *Teach Yourself Basic Maths*. London: Hodder and Stoughton.
Pringle, M. and Cobb, T. (1999) *Making Pupil Data Powerful*. Stafford: Network Educational Press.
Rowntree, D. (1981) *Statistics Without Tears*. London: Penguin.

Index

active learning 5
arithmetic mean 73
averages 73

benchmarking 81
BIDMAS 63
box and whisker diagrams 77
boxplots 77

calculating
　percentages 15
　strategies 26
calculators and percentages 41
charts and graphs 101
combining
　fractions 14
　percentages 18
comments 4
continuous data 70
conversion
　formulas 51
　graph 51
　table 51
converting
　imperial and metric units 51
　percentages 16
　units 49
correlation 118

data 65
decimals 13
discrete and continuous measures 8
discrete data 70

formulas 56
　in spreadsheets 59
fractions 12
frequencies 91
frequency density 108

grouped data 69

histograms 106

imperial units 48
inherent error 47
interquartile range 76
interval scales 71

line graphs 114
line of best fit 118
lower quartile 77

measurements 11
median 73
mental calculations 31

metric measures 48
mode 73
money calculations 40

names and numbers 7
naming scales 71
nature of measures and
    measurement 45

ogive 100
ordering scales 71

paired data 116
PANDA 81
percentage increases/decreases
    18
percentages 14
percentiles 80
performance indicators 119
PICSI 81
pie charts 109
primary data 71

range 76
ratio scales 71
ratios and proportions 20
relative error 47

scales 71
scatter graphs 116
scatterplots 116

secondary data 71
seven-figure summary 83
spread 75
spreadsheets 42
statistical summaries 72
statistical terminology 69
statistics 65
stempots 91
strategies for avoiding common
    errors
        using calculators 42
        with formulas 63
        with fractions, decimals and
            percentages 20
        using measurements 56
        using ratio and proportion
            26
stuck 2
study timetable 5
Système International 48

tables 83
tolerance 47
TTA numeracy skills test 5
types of calculator 37
types of numbers 7

upper quartile 76

weighted scores 60
written calculations 36